T0319017

The orange juice business

M
MARKESTRAT
Value Generation

fea·RP

USP

The orange juice business

A Brazilian perspective

Marcos Fava Neves

Vinícius Gustavo Trombin

(*coordinators*)

Frederico Fonseca Lopes

Rafael Kalaki

Patrícia Milan

Prepared by: MARKESTRAT
Centro de Pesquisa e Projetos em Marketing e Estratégia
[Center for Research and Projects in Marketing and Strategy]

This publication is a translation of 'O retrato da citricultura brasileira', CitrusBR, 2010. Some tables and graphs have been updated with data from 2009/10.

Editing Coordination
Ibiapaba Netto
Prole Gestão de Imagem
Graphic Design
Typodesign
Art Coordination
Typodesign

Art Editor
Alexandre Fedrizzi Ana
Maria Hitomi
Infographics
Ana Emme
Nilson Cardoso

Image Treatment
Marcio Uva
Examination
Arlete Mendes de Souza
Photography
Lau Polinésio
Getty Images
Istockphoto

ISBN: 978-90-8686-181-1
e-ISBN: 978-90-8686-739-4
DOI: 10.3921/978-90-8686-739-4

First published, 2011

Table of contents

Brazil has already achieved high efficiency in its citrus production chain. This efficiency includes everything from certified plant nurseries and seedlings, to the planting and cultivation of oranges, to the production and international distribution of orange juice through integrated bulk cargo systems that include tanker-trucks, port terminals, and dedicated ocean vessels that ship citrus products to consumers in Europe, North America and Asia, with dozens of different specifications and blends for the most diversified applications and unmatched excellence. All with Brazilian competence and know-how. Brazil produces half of the orange juice on the planet, the exports of which generate US$ 1.5-2.5 billion for Brazil every year. In roughly 50 years, the supply chain has generated nearly US$ 60 billion (at today's prices) for Brazil, directly from the world's orange juice consumers.

This wealth is distributed in hundreds of enterprises directly involved in the sector, on thousands of farms, generating over 200,000 direct and indirect jobs, paying taxes, and serving as a driving force for establishments such as Escola Técnica Edson Galvão, in Itapetininga-SP; Qualiciclo Agrícola, in Limeira-SP; Citrograf Mudas Cítricas, in Ipeúna-SP; André Brinquedos, Morada do Sol & FMC, in Araraquara-SP; Restaurante Pantheon and Casa da Cultura, in Matão-SP; Supermercado Alvorada, Itápolis-SP; Fido Construções Metálicas, in Olímpia-SP; Guarnieri Veículos, in Colina-SP; Mercadão dos Tratores, in São José do Rio Preto-SP; Bar Café da Esquina, in Catanduva-SP; Auto Posto Pratão, in Prata-MG; and many other companies located in nearly 400 municipalities in the state of São Paulo dedicated to orange growing, accounting for 80% of Brazil's overall production. In fact, oranges are grown in more than 3,000 municipalities across Brazil.

Oranges compete with other fruits in consumers' choices, and domestic consumption of fresh oranges is increasing, assured by the fact that freshly squeezed orange juice is regularly prepared in homes, bakeries and restaurants throughout the nation, in addition to the market for pasteurized juice, which is produced at factories that operate regionally. Today the domestic market for fresh oranges has become a major consumer of Brazil's total production. More than 100 million boxes of oranges (40.8 kg) – equivalent to approximately 30% of Brazil's production – are consumed by the Brazilian population.

The biggest challenge of this production chain is in exported juice, the destination of the other 70% of Brazil's orange harvest. Orange juice has been losing ground to other juices and beverages, introduced to the market with increasing frequency and steadily gaining market share, whether because they offer fewer calories or lower costs to consumers, or because they represent an opportunity for better profit margins to bottlers and wholesale/retail networks. In the United States, the world's foremost consumer, the demand per capita has decreased by 23% in the last seven years, falling from 23 liters to 17 liters per capita. In the top 14 Western European markets, the decline was from 13 to 12 liters per capita. In Germany, Europe's highest ranking country in orange juice consumption, the decrease was 26%.

So, one option left for orange juice exporters is to look toward emerging nations. But the solution could still be a long way off, because in those countries – with lower per capita income – the categories of nectars and still drinks have the strongest presence on the market. The explanation is the more affordable price to the consumer, because of the low juice content in these beverages. Along with nectars and still drinks, there are the other fruit flavors and other beverage categories, such as sports drinks, teas, coffee-based drinks, flavored milk, and flavored waters, which have experienced higher growth rates in consumption.

Moreover, consolidation in the retail segment is increasing the power of large supermarket chains to push down margins. In Western Europe, 66% of orange juice is sold under retail brand names. In Germany, for example, where the five biggest retailers control 80% of the sale of non-alcoholic beverages, the prices of orange juice to the consumer have varied between €0.59 and €1.00 per liter over the past decade. Striving for higher sales per square foot and greater operational efficiency, these retailers control the availability of space on their shelves, giving preference to those products with higher turnover and that provide greater revenue and profit per square foot of shelf space, thus influencing the consumption patterns in each market.

This strong concentration in retail over the last two decades eventually forced consolidation among bottlers and brands, which are direct buyers of orange juice exported from Brazil. Today, 35 bottlers alone buy 80% of the annual global production of orange juice, the remaining 20% being purchased by 565 bottlers.

Most orange juice buyers are also responsible for bottling and distribution, and the manufacturing infrastructure is also used for other non-alcoholic beverages, for example, other fruit juices, dairy-based beverages, soft drinks, sports drinks, and bottled water. This competition forces them to prioritize bottling of beverages with higher profit margins or those using the cheapest raw materials.

The main objective of this book is to present an X-ray vision of the citrus production chain, providing the reader with a greater understanding of this business, the variables that impact it, as well as its trends and challenges. It gives a thorough outline of the production chain and a number of its principal players.

This book is the result of 12 months of work involving, directly or indirectly, around ten researchers associated with the Center for Research and Projects in Marketing and Strategy (Markestrat), comprised of professors, PhD, Master's, and undergraduate students of the University of São Paulo (USP).

Numerous visits were made to large, medium and small companies in order to collect data and information. There were also numerous debates with the executives of major companies in the industry and discussions at Citrus-BR. Two international trips were also part of the study, one to Nice (France) to participate in the World Juice Congress; and another to the Tetra Pak Worldwide Center for Research & Development and Business Intelligence, in Modena (Italy), for immersion into global data regarding fruit juices. For this we would especially like to thank TetraPak, represented by Paulo Nigro, Eduardo Eisler, Alexandre Carvalho, Bettina Scatamachia, Carol Eckel and Heloisa Rios.

Marcos Fava Neves
Full Professor of Planning and Strategy at FEA/USP, Ribeirão Preto Campus

An overview

Main conclusions of the study

Since 1962, when the first exports began, citrus production has contributed to the development of Brazil in a definitive manner. During this period, the sector has generated US$ 60 billion in exports; in 2010 alone, these exports surpassed US$ 2 billion.

International orange juice prices suffer incredible volatility, capable of fluctuating between US$ 700 and US$ 2,600 per ton within a short period of time.

In 2009, the citrus sector's exports totaled 2.9 million tons, of which 1.129 million tons were in the form of frozen concentrated orange juice (FCOJ), 939,000 tons of not-from-concentrate (NFC) juice, and 851,000 tons of orange by-products.

Consumption

- Three out of every five glasses of orange juice consumed in the world are produced at Brazilian factories.
- Brazil accounts for 50% of global production of orange juice, exports 98% of what it produces, and has achieved an amazing 85% market share worldwide.
- The cost of frozen concentrated orange juice (FCOJ) is only 28% of the shelf price of one liter of juice on the Europe retail market.
- Thirty-five bottlers in Europe purchase 80% of the orange juice exported by Brazil. In the United States, the four largest bottlers retain a 75% market share.
- Orange flavor represents only 0.91% of the worldwide beverage market.
- Orange juice is the world's most-widely consumed fruit-based drink, with a 35% market share among juices.
- Orange flavor has been losing market share to other fruits, with a decline of 1.6% per year. Worldwide demand for orange juice fell by 6% in five years. On the other hand, nectars – which have less soluble solids (fruit sugars) – increased 4% a year.
- In the USA, the world's largest consumer of orange juice, with 38% of the total, consumption decreased by 11.5% in five years. Over ten years, the decline was 24%.
- Florida and São Paulo account for 81% of world production of orange juice. The state of São Paulo alone is responsible for 53% of the total. In the last 15 harvests, global juice production fell by 13%.

Brazil accounts for 50% of global orange juice production, exports 98% of what it produces, and has achieved an amazing 85% market share worldwide.

Florida and São Paulo account for 81% of world production of orange juice. The state of São Paulo alone is responsible for 53% of the total.

Impact on Brazil

> The citrus industry generates – between direct and indirect employment – around 230,000 jobs, and annual payroll of around R$ 676 million

- GDP in the citrus sector is US$ 6.5 billion (2009), with US$ 4.39 billion on the domestic market and US$ 2.15 billion on foreign markets.
- The citrus industry generates – between direct and indirect employment – around 230,000 jobs and an annual payroll of around R$ 676 million.
- Citrus producers invoiced US$ 2 billion in 2009.
- Total sales of all of the links in the citrus production chain were US$ 14.6 billion in 2009.
- Highway concessionaires earned US$ 18.3 million in tolls paid by the citrus industry. The industry spends around US$ 300 million per year on freight costs alone. Roughly one truck per minute passes through toll booths while carrying orange juice from factories in São Paulo to the Port of Santos.
- In the 2009/2010 growing season, Brazilian production totaled 397 million boxes of oranges (40.8 kg each).

> In the 2009/2010 growing season, Brazil's production was 397 million boxes of oranges (40.8 kg each).

Taxes and contributions

- The citrus chain brings in around US$ 189 million in taxes to Brazil.
- The foreign exchange rate is a major enemy of the citrus industry. Taking exports from 2006 to 2009, if the exchange rate were US$ 1.00 to R$ 2.32, the sector would have earned R$ 760 million more per year, representing an additional R$ 2.30 per box processed during this period.
- In 2009, R$ 518 million were paid in tariffs, equivalent to R$ 1.90 per box processed.

> The citrus chain accounts for US$ 189 million in taxes to the Brazilian Government.

Orange groves

- In Brazil, in 2010, there were nearly 165 million trees producing oranges; in Florida there are around 60 million.
- The density of trees per hectare has increased significantly. In 1980 the density was 250 trees per hectare, rising to 357 trees in 1990, 476 trees in 2000, and today there are orange groves with nearly 850 trees per hectare. These days, better seedlings are available, coming from screened nurseries, and nearly 130,000 hectares are already irrigated.
- Roughly 11,000 producers with fewer than 20,000 trees (87% of the total) account for 21% of the existing trees in Brazil's citrus belt. Another 32% of the trees are in the hands of 1,500 producers, who manage between 20,000 and 199,000 trees. In all, 120 producers have more than 200,000 trees, already representing 47% of all orange trees in Brazil.

- The operating cost of production of the industry's orange groves is R$ 7.26 per box. This cost is up from R$ 4.25/box in 2002/2003 (an increase of 70% compare to the current cost of R$ 7.26). Among the costs that increased the most are manpower, which jumped from R$ 0.86 per box to R$ 1.66, and harvesting costs, which climbed from R$ 0.84/box to R$ 2.19/box (a 160% increase). Between 1994 and 2010, minimum wages increased by 628%.
- Pests and diseases were responsible for the eradication of 40 million trees in this decade. The mortality rate jumped from 4% to an alarming 7.5%. These diseases were responsible for losses of nearly 80 million boxes per year. One of the industry's most serious concerns is citrus greening, which is advancing extremely rapidly.

Pests and diseases were responsible for the eradication of 40 million trees in this decade

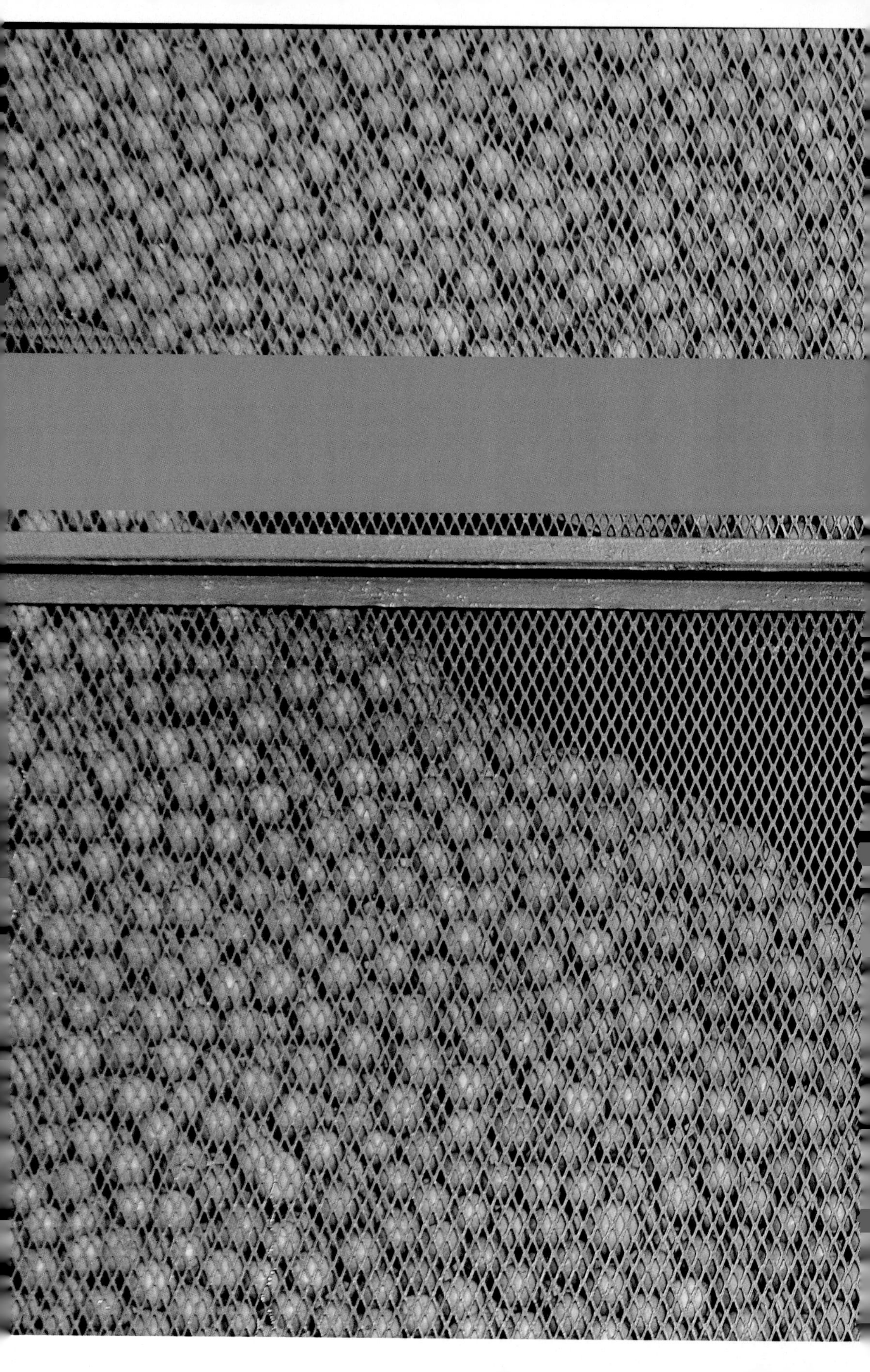

Mapping of the citrus economy

1. The citrus sector in Brazil

Records show that oranges were native to southern Asia, probably China, around 4,000 years ago. Trade and wars between nations helped to expand the cultivation of citrus fruits, and during the Middle Ages, oranges were taken to Europe by the Arabs. In the year 1500, during one of Christopher Columbus' expeditions, citrus seedlings were brought to the Americas (Figure 1).

Introduced to Brazil in the early stages of colonization, orange trees found better conditions to thrive and produce than in their own regions of origin, expanding throughout the territory. The citrus industry was outstanding in several states; however, Brazils first citrus-growing center began to appear around the town of Nova Iguaçu, in the state of Rio de Janeiro, during the 1920s. This orange-growing region would supply the cities of Rio de Janeiro and São Paulo, and usher in exports of oranges to Argentina, England, and other European countries. After this phase, the harvest followed the same paths as coffee, which suffered a significant decline in overall cropland due to severe frost in 1918, the drought in the 1920s, the global financial crisis, and roundworm infestation. Confronted with these problems, the orange harvest shifted to the Paraíba Valley in the state of São Paulo in the 1940s, with the opportunity to replace the coffee harvest in the region of Limeira-SP, later arriving in Araraquara in the early 1950s and in Bebedouro by the end of that decade, gaining more and more ground throughout the new frontiers of the northern and northwestern regions of São Paulo state.

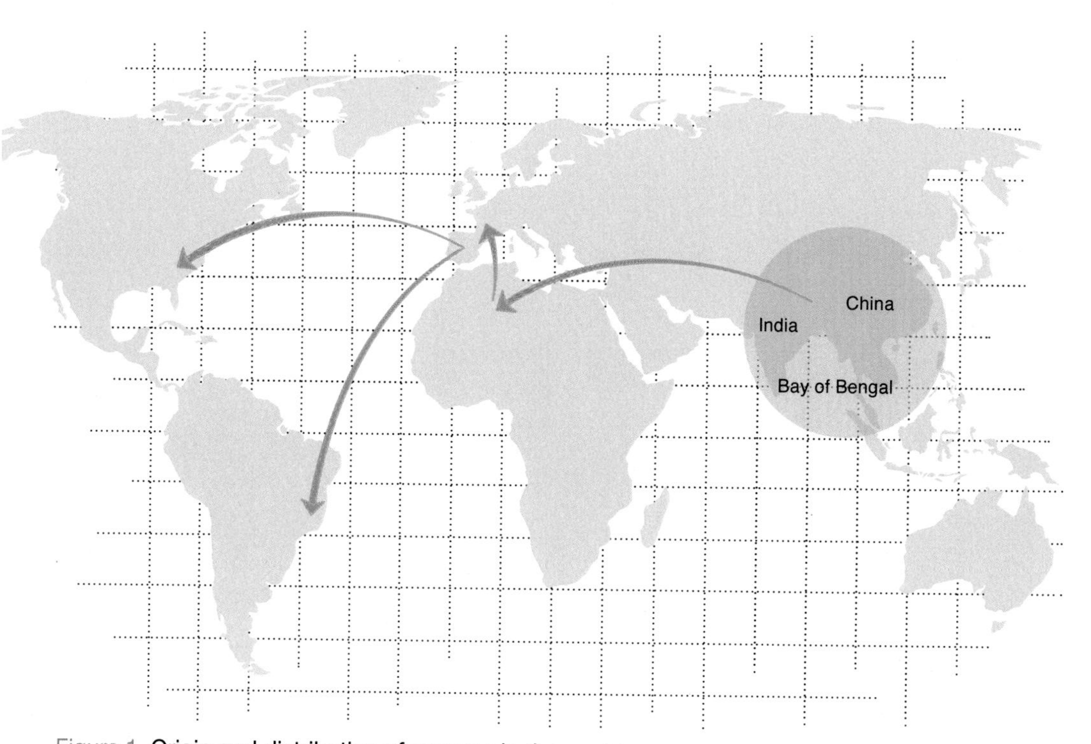

Figure 1. Origin and distribution of oranges in the world.

The development of technology by government agencies associated with the citrus sector assisted in the advancement and consolidation of this activity, allowing citrus-growing to experience a period of rapid expansion and to acquire greater economic importance. From the creation of the citrus- producing center in the early 1920s until 1940, around the onset of World War II, orange production in Brazil had grown more than tenfold. Despite the advances, the industry suffered a critical moment during the war, when the demand for orange exports plummeted.

The recovery of orange exports occurred gradually during the post-war period, but was insufficient to absorb the entire quantity available each growing season. Since the domestic market was underdeveloped, the idea of industrializing the surplus quickly gained a following. In 1959, the first juice concentrate factory was set up in Brazil, and it wasn't long before others started to appear. Currently there are 1,178 juice extracting machines installed in Brazil, of which 1,061 are located in the state of São Paulo, 72 in Southern Brazil, and 45 in Northeastern Brazil (Table 1).

In 1962 a major driving force behind the growth in the Brazilian citrus industry was the frost that struck the orange groves in Florida, USA; until that time the largest producer of orange juice. Brazil, wagering on this sector, worked to fill this gap in the market. In the mid-1960s, the country carried out its first experimental exports of orange juice concentrate.

The consolidation of the Brazilian citrus industry became final when the frost returned to beset Florida in the years 1977, 1981, 1982, 1983, 1985 and 1989, causing losses in the US production of oranges in the respective growing seasons in the order of 23, 30, 38, 52, 16 and 20 million boxes, in addition to a severe decrease in juice content of the fruits due to the freezing of orange pulp and cells. There was also a fall in production in the growing seasons after the frosts, due to the deaths of thousands of trees as a consequence of the freezing temperatures. As a result, exports of Brazilian juice were strengthened and the domestic citrus industry entered a phase of rapid expansion.

The combination of a highly developed citrus sector and a competitive industry helped Brazil become the world's largest producer of oranges in the 1980s, surpassing the United States not only in production but also in citrus-growing technology. In this phase, with a significant drop in production in Florida, the prices of oranges and orange juice reached record highs, helping the Brazilian citrus industry gain more momentum with each growing season. It was a period marked by the rapid planting of new orange groves in São Paulo, with expansion rates of citrus croplands ranging from 12% to 18% annually, and the influx of thousands of new

Table 1. Evolution of the number of extracting machines installed in Brazil per decade.

1970	1980	1990	2000	2010
76	511	815	1,022	1,178

Source: prepared by Markestrat based on data from CitrusBR and FMC.

producers. The increased availability of oranges enabled an increase in orange juice exports and wide availability of oranges for domestic consumption. Citrus fruits, which in many markets were considered a luxury item, began to be consumed by Brazilians of all social classes.

In the 1990s, the Florida citrus-growing sector recovered and shifted its central axis approximately 180 kilometers to the south and southwest of the state, to higher-temperature regions. These new groves were formed using modern irrigation technologies that, besides compensating for the shortage of water, also provide thermal protection for the orange trees in the case of sharp frosts. Production in the State of Florida, which had slumped to 104 million boxes in 1984/85, returned to its peak in 1997/98, with 244 million boxes of oranges.

The resumption of production in Florida and the boom in the growth of citriculture in the Brazilian State of São Paulo, added to the modest rates of growth in consumption, from 2% to 3% per year, and resulted in a surplus of orange juice for the harvests from 1992/93 to 2003/04. The increase in the stocks held by the Brazilian, Florida and European industries led to a devaluing of the juice, both on the futures market and on the physical market, reducing the price of oranges in Florida, Brazil and the Mediterranean region. In this period, the average price for FCOJ (concentrated frozen orange juice) contracts on the New York Stock Exchange was US$ 903 per ton, free of import duty, which represented the equivalent of US$ 3.61 per box of oranges delivered in New York, already processed in the form of concentrated orange juice. In the previous decade, from 1982/83 to 1991/92, the average price for FCOJ on the exchange was US$ 1,583 per ton (see Appendix).

Later, the price of orange juice began to rise again because of three hurricanes in 2004 and one in 2005 that devastated the State of Florida, destroying in each of the respective harvests, 27 million and 39 million boxes of oranges, as well as aiding the dispersion of citrus canker to the heart of the American citrus belt. The reduced supply led to a reduction in the high levels of stock throughout the world, provoking a reaction on the New York Stock Exchange and pushing up prices on the physical market for orange juice in Europe and Asia, which already represented one of the top buyers of national production. The Florida and São Paulo citriculture industry then enjoyed a new cycle of high prices for oranges destined for industrial processing.

In 2009/10, following a significant drop in the price of orange juice, due to the global crisis in 2008, which changed the behavior of consumers, who began to favor cheaper products, an improvement in prices was witnessed, caused by the reduced production in the world's two foremost citrus farming regions. In 2009/10, Brazilian production totaled 397 million boxes of oranges, with exports in 2009 at around 2.9 million tons, with 1,129 thousand tons of FCOJ, 939 thousand tons of NFC (not from concentrated orange juice) and 851 thousand tons of orange by-products.

2. Brazilian GDP versus agricultural GDP

Brazil is, notably, a country focused on agrobusiness. From 1995 to 2008, the sector represented between 24.5% and 28.5% of the country's GDP (Graph 1). Variation in the growth rate for the sector is related to the oscillation of the prices for commodities on the international market and the exchange rate.

Another significant point is the importance of the sector to the trade balance. In 2009, farm product exports were responsible for 42.5% of Brazil's exports, increasing its share by 6% in relation to 2008, despite the fall-out from the financial crisis and the low returns on sales caused by the rise in the exchange rate.

The monetary value of agrobusiness exports in 2009 came to a total of around US$ 65 billion, compared with imports of approximately US$ 10 billion, leading to a trade surplus of US$ 55 billion, underlining the strategic importance of the sector for generating funds. If it were not for agrobusiness, the Brazilian trade balance would go from a surplus of US$ 24.6 billion to a deficit of US$ 30 billion, which would undermine economic stability and the Brazilian real.

It is in this sense that citriculture may be regarded as the generator of a 'clean dollar'. In other words, to export orange juice, a commodity for which the world's top supplier is Brazil,

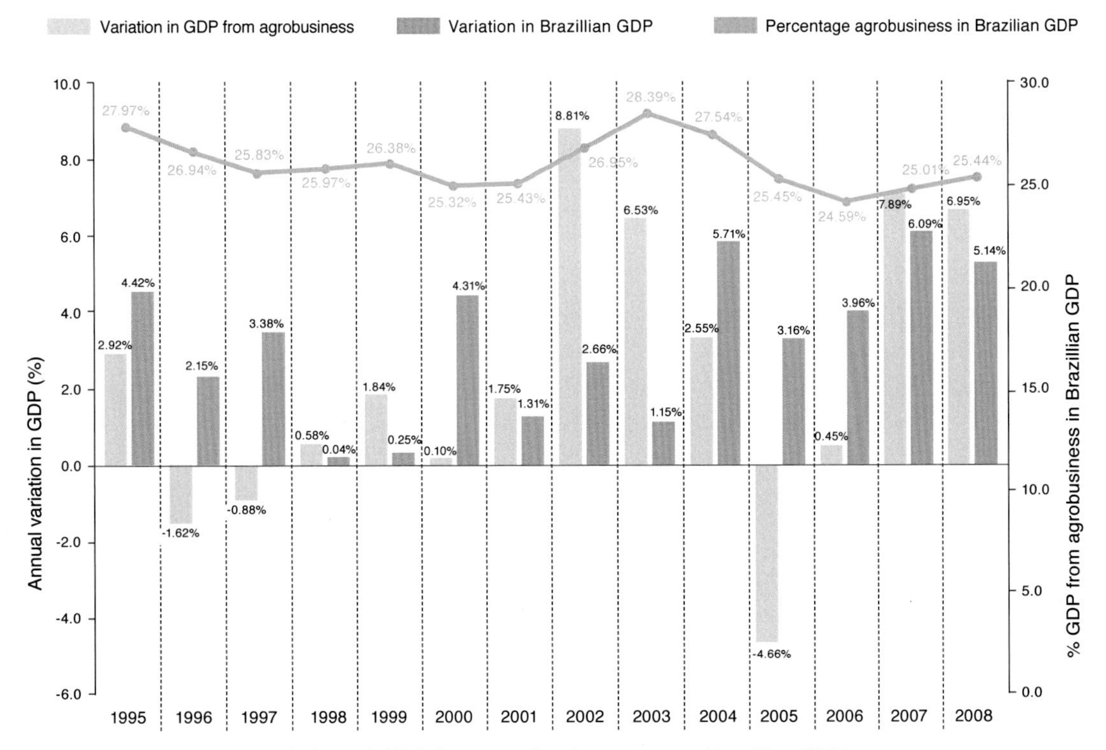

Graph 1. Annual variation of GDP from agrobusiness versus Brazilian GDP.

Source: prepared by Markestrat based on data from the CEPEA.

citrus fresh fruits and other by-products, it is only necessary to import a few supplies, which is not the case for other sectors, such as aviation. In the period from 1962 to 2009, citriculture exported, at 2009 prices, almost US$ 60 billion, bringing in an average of US$ 1.3 billion per year in funds for the country.

3. Brazilian leadership

The prophecy that began with the discovery of Brazil – "everything planted will grow" – seems to be truer with each passing day. This country is the world's largest producer of orange juice, coffee and sugarcane, second in the production of soy and beef, third in chicken and corn and fourth in the production of pork meat.

Along the same lines as the prophecy, another promise recently came to this country – i.e. that Brazil would be a major supplier of foodstuffs to the world. This has also become a reality. The importance of Brazilian production goes beyond territorial borders and is raising its profile in international trade, given that Brazil is responsible for 45% of the world's sugar market and 32% of the global market for coffee (Table 2). However, it is in orange juice that the country demonstrates its leadership. Of every five glasses of orange juice consumed in

Table 2. Position and share of Brazil in the global production and export of agricultural products in 2009.

Product	Production		Exports	
	Position	Share %	Position	Share %
Orange Juice	1st	56	1st	85
Coffee	1st	40	1st	32
Beef	2nd	16	1st	22
Chicken	3rd	15	1st	38
Sugar	1st	22	1st	45
Ethanol	2nd	35	1st	96
Soy (beans)	2nd	27	2nd	39
Soy (middlings)	4th	16	2nd	25
Soy (oil)	4th	17	2nd	21
Corn	3rd	6	2nd	9
Pork	4th	3	4th	12
Cotton	5th	5	4th	9
Milk	6th	6	7th	1

Source: prepared by Markestrat based on GV Agro and USDA (Jan 2010).

the world, three are produced in Brazil (Figure 2). In no other commodity does Brazil have a similar degree of domination.

The strength of the Brazilian orange juice industry is not just in its exports. Its entrepreneurial nature was the driving force, in the 1990s, behind the appearance of the first Brazilian agroindustries to operate on foreign soils, which further strengthened its competitive position on the international scene.

The figures for Brazilian citriculture are impressive. The country is currently responsible for more than half of the world's production of orange juice and exports 98% of its production. The strength of Brazilian orange juice in international trade is a source of pride and has a flavor and respect that are unique to Brazil in the world.

Three out of every five glasses of orange juice consumed in the world

Figure 2. Share of orange juice produced in Brazil in relation to the juice consumed in the world.
Source: prepared by Markestrat based on CitrusBR.

4. Citrus exports

In 2009, exports from the citriculture complex represented a total of 2.15 million tons of products and US$ 1.84 billion in revenues, amounting to around 3% of exports from agrobusiness (Graph 2).

From 2000 to 2009, the revenues obtained rose by 62%, with the share of FCOJ falling from 91% to 71%, due to the increase in exports for the other products within the complex and the start of exports for NFC in 2002, an example of the citriculture industry's reaction to the changes in the habit of consumers who now favor less processed products with a more natural image. In this period there was a 12% reduction in export volumes of FCOJ, despite the 26% increase in the financial value, which was caused by the rise in the prices for orange juice

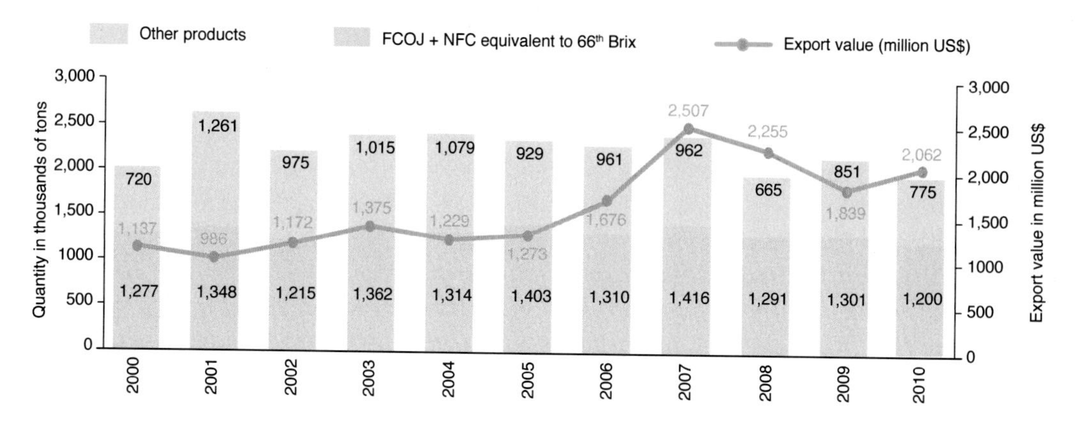

Graph 2. **Quantity and financial value of total exports for the citriculture sector.**
Source: prepared by Markestrat based on Cacex, Banco do Brasil and SECEX/MIDC.

following the hurricanes that hit Florida in 2004 and 2005 and the later reduction in American production of orange juice (Graph 3).

In 2000, exports totaled US$ 85 million (7.5% of the total) revenues in essential oils, D-limonene, terpenes and citrus pulp middlings, these being the by-products of oranges, lemons, limes, tangerines and grapefruit. In 2009, the financial volume rose to US$ 241 million and the share of revenues to 11.3% (Table 3). While the average export price for NFC and FCOJ was, respectively, US$ 337/t and US$ 1,153/t, the export price for essential oil of orange reached U$ 1,966/t, for D-limonene/terpene U$ 1,336/t and for citrus pulp middlings U$ 120/t.

The national consumption of citrus fresh fruits accounts for a significant part of the Brazilian production, although the same cannot be said of the international market, where consumers favor varieties of fresh oranges that are grown in the Mediterranean and

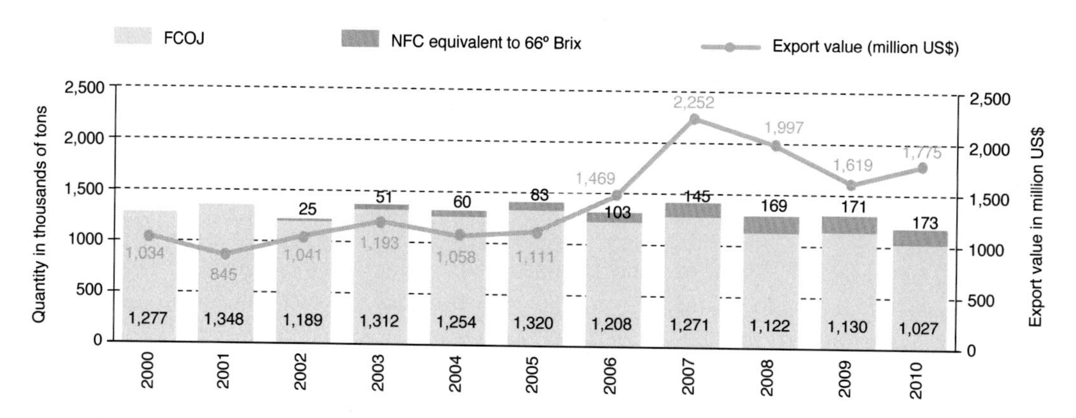

Graph 3. **Evolution in the quantity and financial value of orange juice exports.**
Source: prepared by Markestrat based on Cacex, Banco do Brasil, Siscomex e SECEX/MIDC.

Table 3. Exports for citrus byproducts.

Year	Exports for essential oil of oranges			Exports for D-limonene and citric terpenes (made from oranges, lemons, limes, grapefruits and tangerines)			Exports of citrus pulp pellets (Cpp) citrus pulp middlings		
	Exported volume (tons)	FOB value (US$ total)	Average FOB price (US$/ton)	Exported volume (tons)	FOB value (US$ total)	Average FOB price (US$/ton)	Exported volume (tons)	FOB value (US$ total)	Average FOB price (US$/ton)
2000	17,564	17,177,408	978	38,728	19,258,197	497	557,703	38,307,811	69
2001	26,550	23,325,167	879	41,360	21,174,548	512	1,020,378	61,925,217	61
2002	20,342	38,911,517	1,913	37,927	26,199,384	691	852,682	54,155,697	64
2003	22,852	63,951,368	2,799	36,010	33,883,928	941	858,721	64,974,786	76
2004	27,153	52,375,678	1,929	39,389	29,293,960	744	889,375	64,308,215	72
2005	31,690	57,521,777	1,815	35,379	29,250,376	827	792,959	59,805,413	75
2006	27,845	61,004,192	2,191	41,670	45,673,217	1,096	801,732	72,629,076	91
2007	31,647	70,585,903	2,230	35,316	49,731,956	1,408	799,712	100,033,511	125
2008	30,078	70,892,995	2,357	36,337	54,344,949	1,496	515,021	90,264,450	175
2009	28,408	55,841,684	1,966	35,926	48,009,626	1,336	721,781	86,676,237	120
2010	28,804	74,455,386	2,584	35,278	57,406,131	1,627	663,212	98,809,068	149

Source: prepared by Markestrat based on Cacex, Banco do Brasil, Siscomex and SECEX/MIDC.

California regions, which are their main export centers. This, however, is not the only factor. The phytosanitary trade barriers imposed on the Brazilian orange also make national exports difficult. There is also a need to consider the increase in the production of oranges witnessed in Spain and in a number of African countries. This has resulted in a reduction in the national exports of fresh orange. While in 2001 the exports of fresh oranges reached a total of 3.4 million boxes, the equivalent to US$ 27.5 million, in 2009 just 641 thousand boxes, or US$ 11.3 million were exported (Table 4). This significant drop, as well as all the limitations mentioned, was also caused by the dramatic rise in NFC exports.

Despite the sharp drop in recent years and the low debating on this topic, the financial value of exports for citrus fresh fruits, (orange, lemon/lime, tangerine and grapefruit) is equivalent to around 60% of exports for mango or 45% of exports for grapes. The average price of exported fruits rose between the 2000/01 harvest and the 2009/10 harvest. A 40.8 kg box of fresh oranges rose from US$ 8.00 to US$ 18.00 in this period. The Tahiti lime increased from US$ 11.00 per 27 kg box to US$ 21.00 per box.

Table 4. Exports for the citriculture sector.

Year	FOB value Total exports for the citriculture sector	Volume		
		Processed products		Fresh oranges export volume
	(US$ total)	FCOJ + NFC equivalent to 66° Brix (tons)	Other products and derivatives (tons)	(40.8 kg boxes)
2000	1,136,536,939	1,276,820	719,537	1,846,685
2001	985,955,684	1,348,196	1,260,641	3,421,150
2002	1,171,943,582	1,214,833	975,382	989,565
2003	1,374,742,812	1,362,331	1,014,696	1,667,050
2004	1,229,337,711	1,314,301	1,079,043	2,210,043
2005	1,272,929,023	1,403,468	929,029	751,326
2006	1,676,319,828	1,310,309	961,471	1,228,934
2007	2,506,795,880	1,415,523	961,577	1,219,331
2008	2,255,379,787	1,291,299	665,213	937,678
2009	1,838,972,527	1,300,554	851,411	641,795
2010	2,061,547,290	1,199,929	775,048	927,005

Source: prepared by Markestrat based on Cacex, Banco do Brasil, Siscomex and SECEX/MIDC.

5. Destinations of exports

Europe stands out as the foremost destination for exports of Brazilian orange juice (Graph 4). In the 2009/10 harvest, 71% of the amount exported entered via the Netherlands and Belgium, re-exporters to the other European countries. Added to the exports made to the United States, these two destinations account for more than 90% of Brazilian orange juice exports.

However, in the last decade, Brazil has managed to diversify the markets in which it trades. In the 2009/10 harvest, Brazil exported orange juice to 70 different countries, of which 12 received NFC (Graph 5 and Table 5 and 6). This demonstrated the capacity for innovation in the industry by redirecting exports to non-saturated markets, finding new channels for offloading national production.

It is worth underlining that the United States is not an especially significant import market for citrus fresh fruits from Brazil. Around 80% of the volume of citrus fruits traded on the international market are destined for Europe. However, Saudi Arabia and the Arab Emirates are beginning to gain importance as destination markets. Together they purchased 8% of the volume exported by Brazil in the 2009/10 harvest.

If the old continent is seen as being a traditional customer for Brazil, the Middle Eastern countries, given the purchasing power and the habit their populations have of not consuming alcoholic beverages, and Asia, with its high population, represent potential markets for growth in the consumption of citrus products. Most countries in these regions, however, currently consume the juice highly diluted, in the form of a freshly squeezed orange juice.

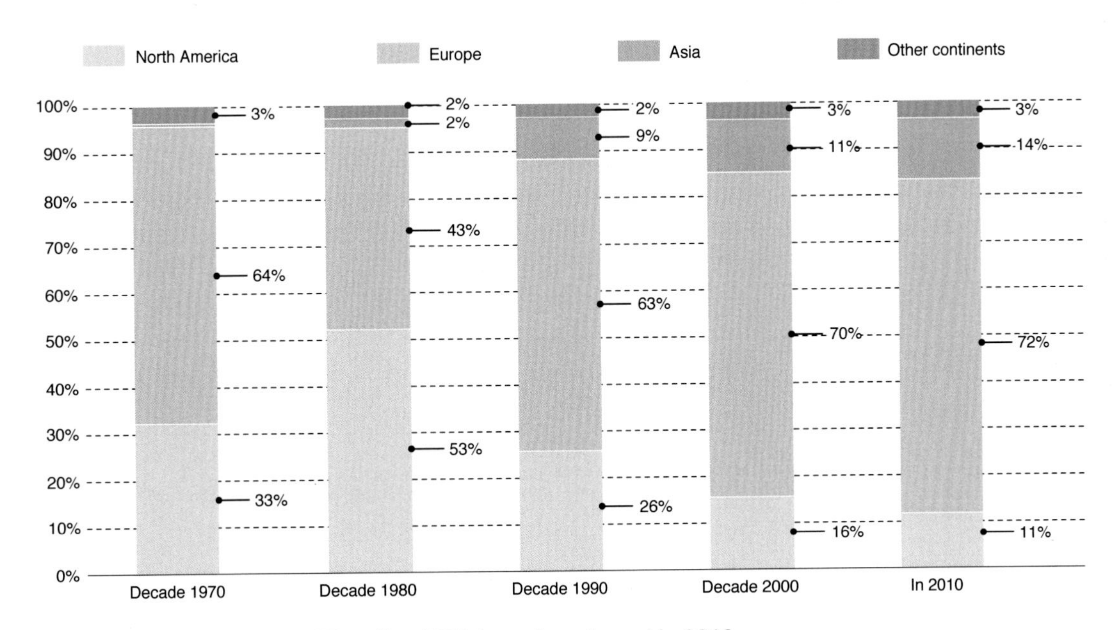

Graph 4. **Destination of Brazilian FCOJ per decade and in 2010.**

Source: prepared by Markestrat based on Cacex, Banco do Brasil, Siscomex and SECEX/MIDC.

5. Destinations of exports

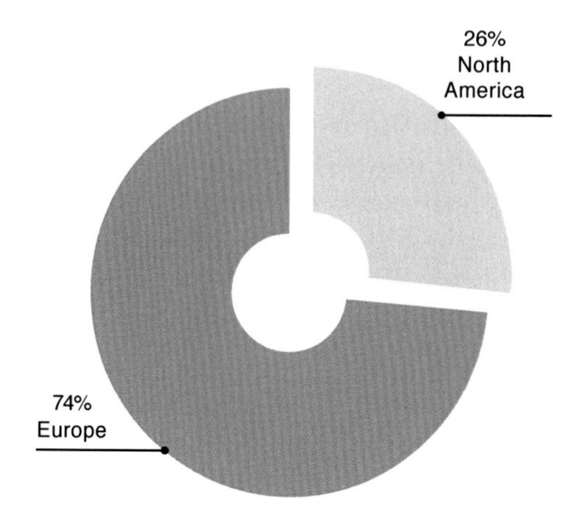

Graph 5. Destination of Brazilian NFC in the decade of 2000.

Source: prepared by Markestrat based on Cacex, Banco do Brasil, Siscomex and SECEX/MIDC.

Table 5. Share of FCOJ and NFC in Brazilian orange juice exports.

Year	FCOJ		NFC equivalent to 66° Brix		Total	
	Tons 66° Brix	Share	Tons 66° Brix	Share	Tons 66° Brix	Share
2000	1,276,820	100%	-	0%	1,276,820	100%
2001	1,348,196	100%	-	0%	1,348,196	100%
2002	1,189,463	98%	25,369	2%	1,214,833	100%
2003	1,311,682	96%	50,650	4%	1,362,331	100%
2004	1,254,355	95%	59,946	5%	1,314,301	100%
2005	1,320,328	94%	83,140	6%	1,403,468	100%
2006	1,207,701	92%	102,607	8%	1,310,309	100%
2007	1,270,927	90%	144,596	10%	1,415,523	100%
2008	1,121,829	87%	169,470	13%	1,291,299	100%
2009	1,129,747	87%	170,808	13%	1,300,554	100%
2010	1,027,104	86%	172,826	14%	1,199,929	100%

Source: prepared by Markestrat based on Cacex, Banco do Brasil, Siscomex and SECEX/MIDC.

Table 6. Share per buyer market in destination of Brazilian orange juice exports.

Year	North America		Europe		Asia		Other continents		Total
	Tons Brix original	Share	Tons Brix original	Share	Tons Brix original	Share	Tons Brix original	Share	Tons Brix original 66° Brix
Exports of FCOJ (frozen concentrated orange juice)									
2000	261,356	20%	848,589	66%	122,715	10%	44,159	3%	1,276,820
2001	190,008	14%	973,673	72%	137,613	10%	46,901	3%	1,348,196
2002	194,872	16%	847,686	71%	119,843	10%	27,061	2%	1,189,463
2003	228,953	17%	913,515	70%	141,238	11%	27,975	2%	1,311,682
2004	147,143	12%	928,820	74%	142,948	11%	35,445	3%	1,254,355
2005	204,360	15%	895,715	68%	185,778	14%	34,474	3%	1,320,328
2006	183,541	15%	831,750	69%	153,827	13%	38,584	3%	1,207,701
2007	252,434	20%	844,820	66%	142,085	11%	31,588	2%	1,270,927
2008	159,254	14%	807,757	72%	114,430	10%	40,387	4%	1,121,829
2009	141,505	13%	797,819	71%	150,213	13%	40,210	4%	1,129,747
2010	108,931	11%	743,448	72%	141,650	14%	33,075	3%	1,027,104
Exports of NFC (not from concentrated orange juice)									
2002	29,644	21%	109,437	78%	1	0.0%	450	0.3%	139,532
2003	73,564	26%	204,610	73%	1	0.0%	397	0.1%	278,572
2004	78,630	24%	245,630	75%	214	0.1%	5,229	1.6%	329,703
2005	83,033	18%	371,000	81%	156	0.0%	3,083	0.7%	457,272
2006	134,478	24%	428,134	76%	767	0.1%	962	0.2%	564,341
2007	256,590	32%	536,831	68%	1,292	0.2%	565	0.1%	795,278
2008	206,670	22%	722,581	78%	2,156	0.2%	679	0.1%	932,086
2009	280,112	30%	658,062	70%	666	0.1%	601	0.1%	939,442
2010	262,070	28%	688,162	72%	293	0.0%	17	0.0%	950,541

Source: prepared by Markestrat based on Cacex, Banco do Brasil, Siscomex and SECEX/MIDC.

6. Tariff barriers

Brazilian orange juice comes up against tax barriers that reduce its competitiveness on the international market. To enter Europe, Brazilian juice is taxed at 12.2% (Table 7). In contrast no tax is levied on juices coming from the Caribbean, North Africa and Mexico[1]. In the United States, the tax paid on FCOJ is US$ 415/ton (Table 7), which incurs additional costs for American consumers, while NFC is taxed at US$ 42/ton. In contrast no tax is levied on imports coming from Central America, Mexico and the Caribbean. Other countries that also impose taxes on Brazilian orange juice are Japan, South Korea, China and Australia.

With the exception of the United States, where the tax is a fixed amount according to volume, the other countries consider the financial selling price. This being so, the higher the price of the orange juice, the higher the rate of tax paid by Brazil. This dynamic boosts the effect of price rises for the price on the supermarket shelves, diminishing the competitiveness of the orange flavor option in relation to juices made from other fruits, such as apples, pears, raspberries and strawberries, which are largely produced in the regions in which they are consumed, and are therefore exempt from tax barriers in their markets.

Table 7. Amount of import taxes for Brazilian orange juice and estimated amount paid in 2009 by the citrus farming sector.

Country/region	Import tax rate	Volume exported in 2009 (in tons)		Estimated import tax paid in 2009[a] (in US$ million)	% of the financial value of taxes paid by Brazil in 2009 to the importing country
		FCOJX	NFC		
Europe	12.20%	797,819	658,062	166.7	64%
United States	FCOJ: US$ 415 /ton NFC: US$ 42 /ton	106,505	258,112	55.0	21%
Japan	25.50%	71,351	-	23.7	9%
South Korea	54%	12,241	-	8.6	3%
China	7.5% for juice at below -18 °C and 30% for juices at temperatures higher than -18 °C	48,900	-	4.8	2%
Australia	5%	26,220	-	1.7	1%
Other destinations[b]	Exempt	66,712	22,000	Exempt	-
Total	-	1,129,748	938,174	260.4	100%

[a] Average sale price considered for calculating the FCOJ: US$ 1,300.00 and NFC: US$ 500.00.

[b] Mexico is exempt until it attains a volume of 30 thousand tons per year. However, current Mexican exports to Europe do not attain this quantity and are, therefore, exempt from tax.

Source: prepared by Markestrat based on data from SECEX.

In 2009, Brazilian exports of orange juice were taxed at around US$ 260.4 million (Table 7). In comparative terms, the amount paid is almost US$ 1.00 per box processed in the citrus belt or US$ 1,600 per worker involved in the growing and processing of the oranges, considering the fixed and temporary workers in the 2009/10 harvest. The elimination of these high taxes could bring better pay rates to the entire productive chain, due to the increase in external resources that would enter the country or due to the possibility of an increase in demand from the international market, following the reduction in the end cost of the product to the consumer.

7. Phytosanitary barriers and technical requirements

The main destination markets for citrus products, Europe and the United States, are countries with different market legislation. Brazilian exporters are required to comply with a number of demands concerning phytosanitary issues, packaging, consistency in product quality and regularity of delivery.

The European Union demands compliance with local legislation and the Codex Alimentarius, a set of rules accepted throughout the world regarding the production of foodstuffs and food safety. Respect must also be afforded to the legislation of the exporting market, covering the general laws for exporting foodstuffs, the specific laws for fruit juices, the presence of contaminants, pesticides and requirements relating to certificates, especially for organic goods.

Besides the tax barriers, the sector has to deal with technical requirements from the importing nations, which increase the cost of supplying orange juice and that represent real barriers and not just tax barriers. In the European Union, for example, the list of acceptable pesticides differs to that for Brazil with respect to a number of products that are crucial to citrus fruit production.

In China, there is a difference in the taxes on imports in accordance with the temperature of the juice: 7.5% for juice at below -18 °C and 30% for juices stored at temperatures higher than this. This is a tax barrier that raises the end cost of the orange juice for the Chinese consumer, since it discourages the adoption of a tanker transport system (transported at temperatures between -8 °C and -10 °C) the logistics cost of which is much more competitive than for juice transported in drums. Also in China, the maximum levels of microbiological contamination are 25 times stricter than those for Europe and up to 50 higher than the levels considered acceptable on the North American market.

However, despite the recognized competitiveness of the national citrus farming chain, these requirements and their different levels of tolerance, which vary in accordance with the prospects for supply and demand, end up restricting exports. In citrus farming, some of the non-tax barriers that make it difficult for fresh fruit to enter the European Union are: the application of phytosanitary restrictions for blackspot and citrus canker and the imposition of maximum limits for residual pesticides. The United States prohibit imports of citrus fruits produced in any part of Brazil due to the Mediterranean fruit fly.

Brazilian exporters are constantly attentive to the preferences of its importing markets. In the case of Europe, the main market for Brazilian orange juice, the main requirements of the European importers concern safety (consumer health, levels of contaminants, pesticide residues), quality (sensory appeal and compliance with technical specifications), authenticity (adulteration, compliance with legislation), traceability (identity of the product in the chain of fruit juices, ease of finding the source of possible problems), and perception of the consumers (product image, origin). With regard to legal requirements, the local legislation must be observed, the Codex Alimentarius, legislation for the export market (legislation pertaining to foodstuffs, juices, contaminants, additives, pesticides, allergenics, organics).

The orange juice encountered by consumers on the supermarket shelves differs greatly to fit the consumer habits of each market (preference for more diluted juice, more bitter juice, etc) and the individual consumers (juice with pulp, without pulp, with added sugar etc.). However, Brazilian industry has to supply its European customers with a "homogenized" product, in strict compliance with the technical specifications, for blendings and adaptations to be produced on the consumer markets, representing a major cost for the industry and, consequently to the productive chain in Brazil.

8. Exchange rate

The real exchange rate, adjusted according to the *inflação*[1] in Brazil and in the United States, seeks to reflect the purchasing power of the national currency in relation to the US dollar, as well as the competitiveness of the country on the international market. A nominal exchange rate devalued in relation to the real exchange rate, such as was the case from the end of 1998 to the start of 2005, stimulates exports (Graph 6).

On the other hand, an increased value in the nominal exchange rate generates fewer revenues in Brazilian Reais for the country, as can be seen in Brazilian exports of orange juice from 2006 to 2009 (Graph 6). In this period, the revenues obtained came to approximately US$ 8.3 billion, the equivalent of R$ 16.3 billion, considering the nominal exchange rate (R$ 1.99/US$). If the real exchange rate were used (R$ 2.32/US$), the amount would be R$ 19.1 billion, a difference of R$ 2.8 billion more in revenues for the period from 2006 to 2009, approximately R$ 706 million per year, or R$ 2.30 per box of oranges processed by the industry (Graph 7).

If on the one hand the devaluation of the US dollar has undermined revenues in Brazilian Reais for the Brazilian orange juice exporter, it has benefited the European importer on the other. The increased value of the Euro in relation to the Dollar has raised its purchasing power in such a way that this commodity has become relatively cheaper. This, allied with the stability of the price of orange juice on supermarket shelves, has allowed a transfer of income from the first links in the productive chain, located in Brazil, to the end links located in Europe.

Despite the unfavorable exchange rate for exporters, the citrus farming sector has maintained export levels for its complex of products due to the importance of Brazil as a

[1] Index used by the Brazilian Central Bank (BACEN): INPC for Brazil and external IPCs for the United States.

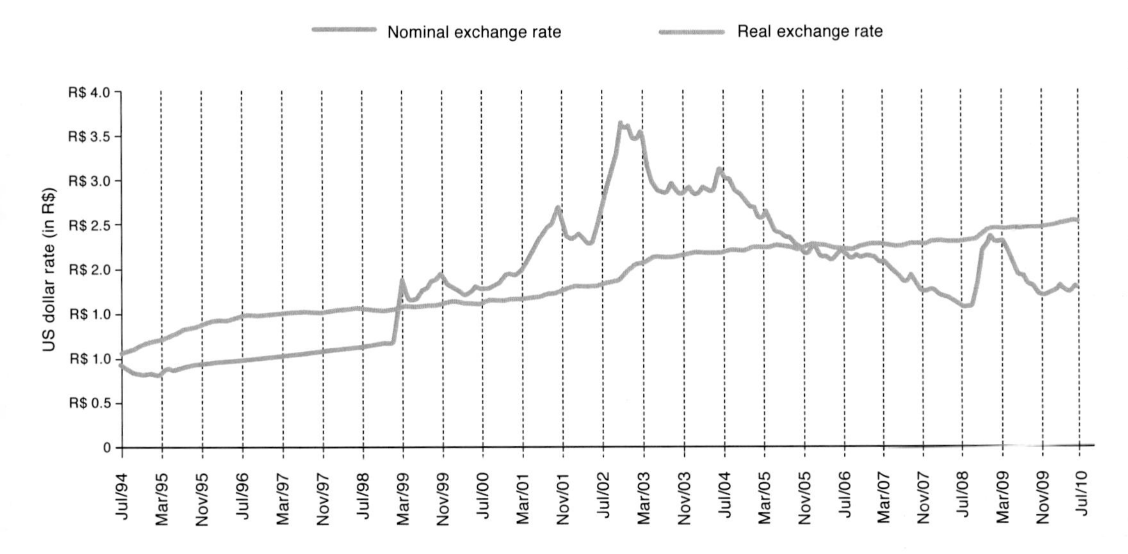

Graph 6. **Nominal exchange rate versus real exchange rate.**
Source: prepared by Markestrat, based on Brazilian Central Bank (BACEN).

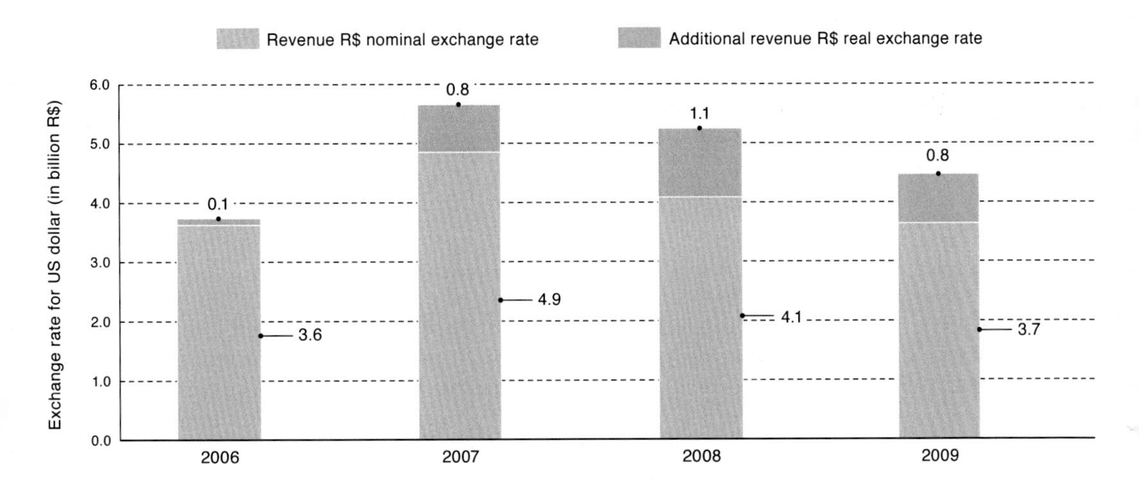

Graph 7. **Estimated revenues from citrus exports: nominal exchange rate and real exchange rate.**
Source: prepared by Markestrat, based on data from the Brazilian Central Bank (BACEN).

world supplier, especially for orange juice. Many other sectors of Brazilian industry, such as the footwear industry, experienced greater difficulties in resisting the pressure brought about by the increased value of the Brazilian Real.

In a nutshell, today, the exchange rate is one of the most punishing and value-detracting factors in the production chain.

Citrus fruit production mapping

9. Evolution of global orange production

In the last six years, growth in the area of land planted with citrus fruits throughout the world was approximately 17%, reaching a global total of around 7.63 million hectares (Graph 8).

The area of land for citrus farming is the second largest in relation to other fruits, losing only to the production of bananas (10.2 million hectares). Among the citrus products the area destined for oranges represents around 55%, thereby consolidating this fruit as the most important player in citriculture (Table 8). However, this percentage has been higher; in 1979 it was 65%. In the citrus farming centers of California, the Mediterranean and Asia, year after year, the orange has been losing ground to other citrus fruits, especially tangerines and mandarins which, because they are easy to peel and eat, have become more valued by consumers of fresh fruit. In global terms, data for the last ten years shows that growth in the area destined for oranges stands at 13%, while that for tangerines has grown by 30%.

Brazilian leadership in the production of oranges began with the 1981/82 harvest, when national production surpassed that of the United States, in the wake of a sequence of frosts that hit Florida, which is the primary orange farming region in the USA. Since then, Brazilian production has practically doubled and the United States have maintained their place as the second largest producer of oranges. However, with each passing year their production falls and they currently achieve less than half the Brazilian production. Next in order of ranking come China, India, Mexico, Egypt, Spain, Indonesia, Iran and Pakistan which jointly produce practically the same volume as Brazil and the USA together. After that, there are another 111 countries jointly producing practically the same amount as that produced solely by Brazil (Graph 9 and Table 9).

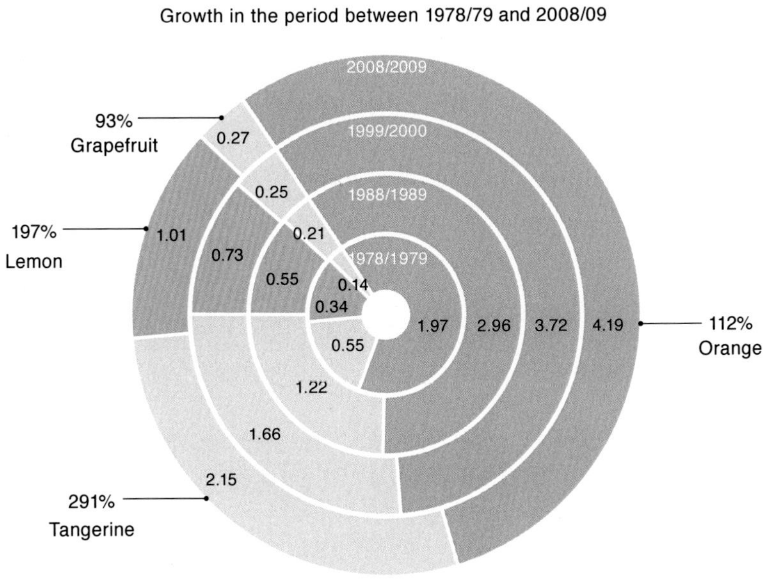

Growth in the period between 1978/79 and 2008/09

2008/2009
1999/2000
1988/1989
1978/1979

93% Grapefruit: 0.14, 0.21, 0.25, 0.27
197% Lemon: 0.34, 0.55, 0.73, 1.01
291% Tangerine: 0.55, 1.22, 1.66, 2.15
112% Orange: 1.97, 2.96, 3.72, 4.19

Graph 8. Evolution of area planted with citros in the world.

Source: prepared by Markestrat based on data from FAO, USDA and CitrusBR.

Table 8. Comparison between Brazil and the world in size of area, production and productivity of selected cultures.

Culture	Planted area (thousand hectares)						Production (thousand tons)	
	World		Brazil		Share of Brazil in relation to the rest of the world		World	
	1998/99	2008/09	1998/99	2008/09	1998/99	2008/09	1998/99	2008/09
Sugar cane	19,318	24,375	4,986	8,140	26%	33%	1,275,520	1,743,093
Corn	139,854	162,059	10,585	14,445	8%	9%	624,413	831,895
Wheat	222,846	226,024	1,409	2,364	1%	1%	596,633	691,863
Rice	151,696	158,955	3,062	2,851	2%	2%	579,187	685,013
Fruits	49,984	56,214	2,319	2,256	5%	4%	475,403	646,450
Banana	8,994	10,208	518	513	6%	5%	92,158	125,049
Citrus	6,352	7,622	1,123	936	18%	12%	87,482	112,819
Orange	3,720	4,189	1,019	837	27%	20%	56,465	69,021
Tangerine	1,655	2,154	57	54	3%	3%	15,713	25,442
Lemon/lime	730	1,013	46	44	6%	4%	10,746	13,439
Grapefruit	248	265	2	2	1%	1%	4,557	4,917
Grape	7,215	7,408	61	80	1%	1%	57,033	67,709
Apple	5,783	4,848	26	38	0.1%	1%	56,654	69,604
Watermelon	2,790	3,753	77	88	3%	2%	60,588	99,194
Other fruits	18,851	22,375	513	601	3%	3%	121,489	172,075
Soybean	70,982	96,870	13,304	21,057	19%	22%	160,135	230,953
Green coffee	10,060	9,722	2,082	2,170	21%	22%	6,647	8,235
Others	479,877	515,494	10,172	14,377	2%	3%	1,993,873	2,432,346
Total	1,144,617	1,249,714	47,919	67,660	4%	5%	5,711,811	7,269,847

Source: USDA, FAO, IBGE, CONAB, CitrusBR.

				Yield (kg/ha)					
Brazil		Share of Brazil in relation to the rest of the world		World		Brazil		Comparison between Brazil and the rest of the world	
1998/99	2008/09	1998/99	2008/09	1998/99	2008/09	1998/99	2008/09	1998/99	2008/09
345,255	645,300	27%	37%	66,028	71,510	69,247	79,274	5%	11%
29,602	58,933	5%	7%	4,465	5,133	2,796	4,080	-37%	-21%
2,270	6,027	0%	1%	2,677	3,061	1,611	2,550	-40%	-17%
7,716	12,061	1%	2%	3,818	4,309	2,520	4,231	-34%	-2%
37,998	40,192	8%	6%	9,511	11,500	16,389	17,813	72%	55%
7,119	6,998	8%	6%	10,247	12,250	13,732	13,639	34%	11%
19,463	19,807	22%	18%	13,773	14,802	17,333	21,152	26%	43%
18,360	17,422	33%	25%	15,180	16,477	18,025	20,825	19%	26%
625	1,273	4%	5%	9,496	11,810	11,063	23,621	17%	100%
428	1,040	4%	8%	14,729	13,262	9,338	23,678	-37%	79%
50	72	1%	1%	18,405	18,523	25,000	36,000	36%	94%
772	1,421	1%	2%	7,905	9,140	12,709	17,780	61%	95%
791	1,124	1%	2%	9,797	14,358	30,072	29,527	207%	106%
1,796	1,995	3%	2%	21,714	26,434	23,315	22,624	7%	-14%
8,057	8,846	7%	5%	6,445	7,690	15,701	14,728	144%	92%
31,307	59,242	20%	26%	2,256	2,384	2,353	2,813	4%	18%
2,164	2,760	33%	34%	661	847	1,040	1,272	57%	50%
8,883	16,926	0.1%	1%	4,155	4,718	873	1,177	-79%	-75%
465,196	841,442	8%	12%	4,990	5,817	9,708	12,436	95%	114%

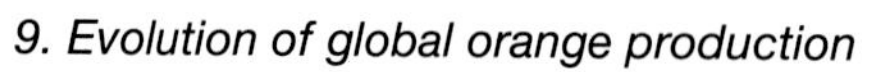

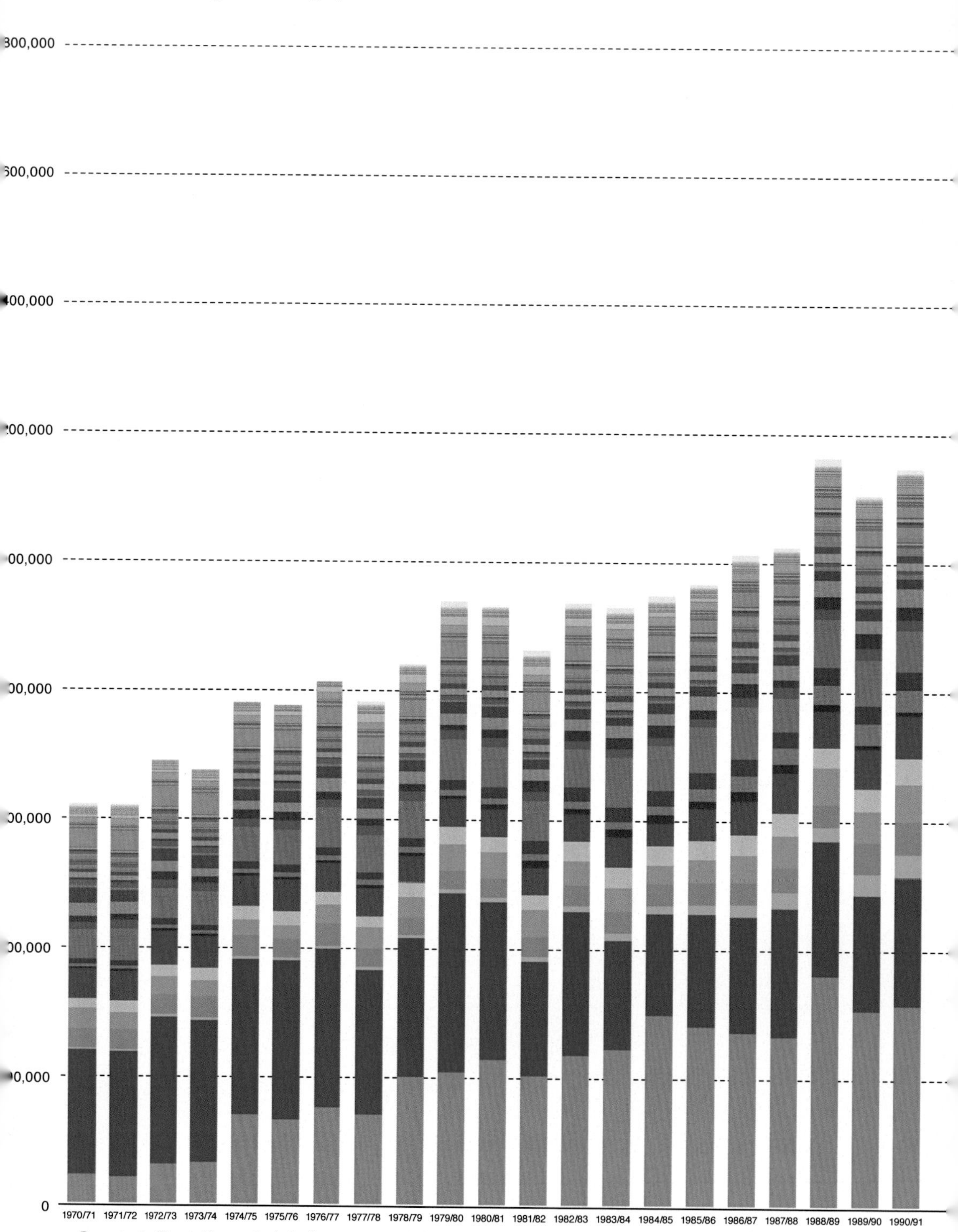

Graph 9. Evolution of global orange production.

Source: prepared by Markestrat based on USDA, FAO, IBGE, CONAB and CitrusBR.

9. Evolution of global orange production

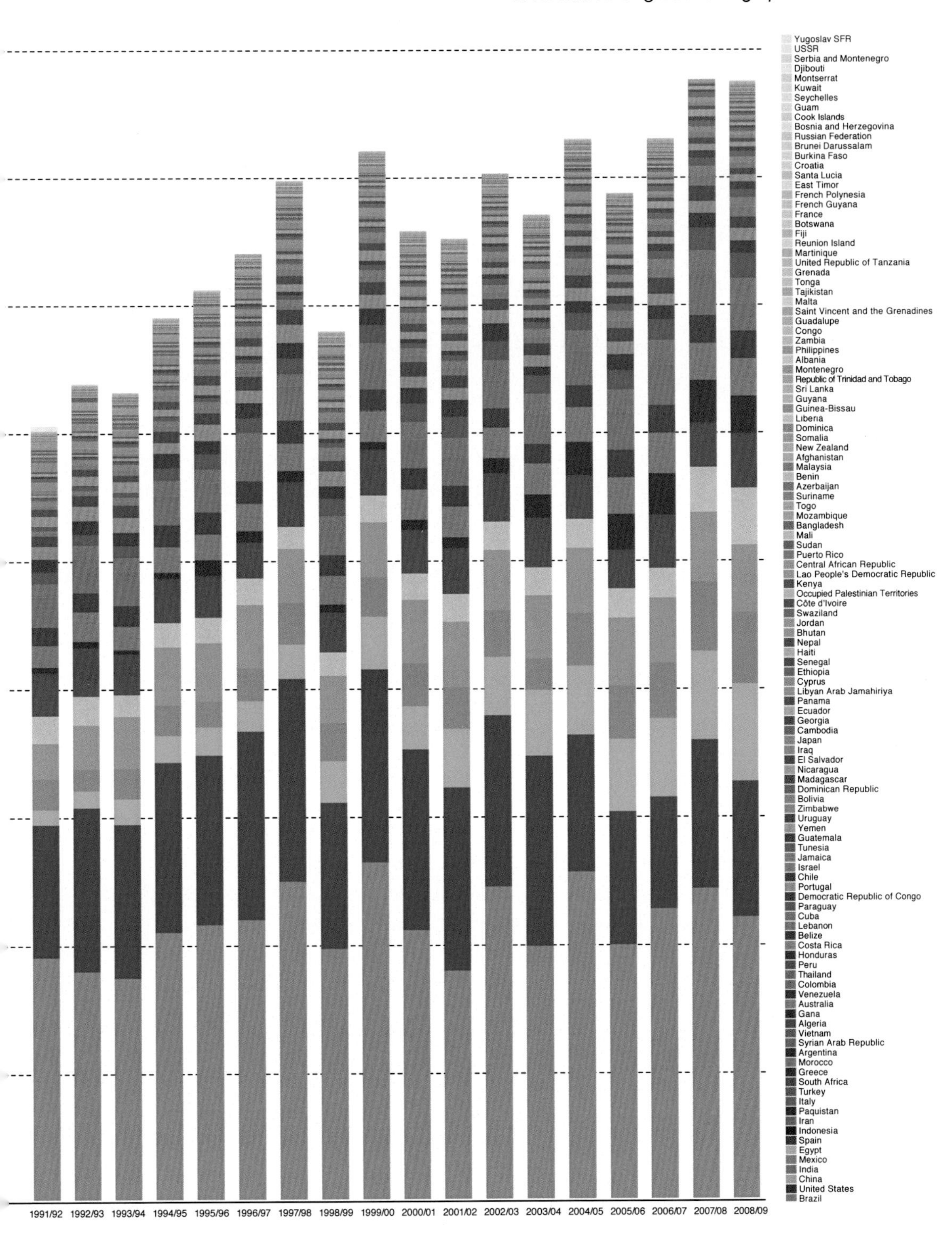

Table 9. Evolution of global orange production in million of boxes.

Country	1970/71	1975/76	1980/81	1985/86	1990/91	1995/96	2000/01	2005/06	2008/09
Total	600,032	756,356	900,681	932,366	1,108,786	1,376,140	1,465,352	1,523,894	1,691,680
Brazil	43,480	127,451	219,020	269,975	302,990	416,005	407,562	384,930	427,010
United States	187,451	238,824	236,789	169,191	193,701	256,225	273,015	200,882	204,510
China	2,307	4,115	6,982	14,142	33,680	42,328	64,583	109,069	147,059
India	29,412	28,186	28,701	33,088	49,265	39,098	65,556	81,225	107,762
Mexico	*30,752*	20,466	39,216	34,559	56,373	87,990	95,221	101,887	101,471
Egypt	13,897	20,980	22,574	28,627	38,578	38,113	39,461	44,118	85,784
Spain	44,926	47,304	41,642	47,598	63,480	63,064	65,882	58,235	82,525
Indonesia	2,623	4,103	7,623	11,893	6,217	24,623	15,786	54,265	56,926
Iran	4,363	6,716	8,824	18,117	32,691	38,135	45,185	55,226	56,373
Pakistan	7,647	11,520	16,770	24,608	27,608	33,618	32,549	42,181	42,181
Italy	32,377	38,725	42,574	55,319	43,137	44,118	44,118	55,417	41,495
Turkey	10,907	13,235	17,034	12,377	18,015	20,637	26,225	35,417	38,235
South Africa	11,054	13,824	13,848	12,181	15,882	22,794	27,426	28,603	37,402
Greece	9,657	13,064	12,917	13,578	20,074	20,539	23,922	24,926	19,363
Morroco	*19,299*	14,314	17,034	20,613	27,034	24,828	16,985	19,216	19,363
Argentina	24,265	18,211	*17,255*	15,270	14,706	17,230	22,377	20,588	17,157
Syrian Arab Republic	105	362	857	944	4,191	7,429	9,977	11,103	14,778
Vietnam	1,838	1,961	2,042	2,727	2,923	9,299	10,458	14,738	14,730
Algeria	8,833	8,250	6,879	4,164	4,506	5,557	7,343	10,668	12,032
Gana	3,162	3,686	3,779	1,961	1,225	4,902	7,353	12,255	11,765
Australia	7,892	8,848	10,441	12,108	11,887	14,436	10,711	11,520	10,539
Venezuela	5,025	6,512	8,115	9,073	10,614	14,546	12,176	9,177	9,555
Colombia	2,230	2,877	5,517	5,882	3,945	9,694	8,649	2,357	8,652
Thailand	3,370	3,676	4,289	5,841	7,353	7,966	7,966	8,578	8,578
Peru	5,968	4,369	3,078	3,565	3,986	5,595	6,268	8,198	8,438
Honduras	448	686	892	1,225	885	2,083	2,941	6,985	7,108
Costa Rica	1,449	1,667	1,838	1,961	2,713	3,676	9,926	9,380	6,814
Belize	558	801	1,110	1,044	1,697	3,182	5,231	6,268	5,870
Lebanon	4,386	5,392	5,392	5,882	6,863	3,431	3,735	5,775	5,605
Cuba	3,160	3,110	7,453	9,961	14,751	7,015	11,532	9,546	4,912
Paraguay	4,338	4,061	4,412	5,175	4,285	4,204	4,946	7,395	4,657
Democratic Republic of Congo	3,157	3,542	3,431	3,676	4,289	4,902	4,534	4,418	4,434
Portugal	2,397	2,941	2,721	3,407	4,328	5,126	6,263	5,363	4,343
Chile	1,054	1,150	1,451	1,838	2,382	2,647	2,377	3,480	3,799
Israel	25,686	23,725	18,529	16,789	13,897	10,784	5,343	3,529	3,799

Table 9. Continued.

Country	1970/71	1975/76	1980/81	1985/86	1990/91	1995/96	2000/01	2005/06	2008/09
Jamaica	1,597	1,001	1,058	1,093	1,557	3,452	3,382	3,431	3,480
Tunisia	1,571	2,169	2,257	2,990	3,020	2,475	2,819	2,485	3,431
Guatemala	-	-	1,127	1,544	1,990	1,960	2,557	3,386	3,300
Yemen	-	-	66	68	238	972	3,884	2,058	3,217
Uruguay	1,033	1,005	1,510	1,838	2,876	3,104	2,675	4,326	3,160
Zimbabwe	515	735	711	907	1,520	1,716	2,402	2,328	2,279
Bolivia	1,324	1,643	2,101	938	1,923	2,266	2,696	2,149	2,246
Dominican Republic	1,566	1,593	1,740	1,497	1,480	1,900	3,221	2,470	2,214
Madagascar	1,391	1,995	1,443	1,527	2,083	2,010	2,034	2,120	2,206
Nicaragua	1,100	1,225	1,275	1,495	1,618	1,765	1,593	1,765	2,083
El Salvador	949	1,144	2,429	2,452	2,625	2,235	913	1,380	1,789
Iraq	735	1,471	2,574	3,848	4,412	7,794	6,618	1,936	1,789
Japan	6,417	9,480	9,841	8,118	5,385	3,343	2,549	1,831	1,593
Cambodia	1,005	735	539	882	1,054	1,225	1,544	1,544	1,544
Georgia	-	-	-	-	-	2,892	980	2,999	1,353
Ecuador	3,748	6,127	13,076	5,655	1,889	2,111	3,661	1,936	1,346
Panama	1,338	1,529	779	855	645	659	674	1,028	1,134
Libyan Arab Jamahiriya	416	622	1,268	1,716	2,230	1,103	1,042	1,087	1,103
Cyprus	2,415	797	772	1,115	1,544	1,348	1,047	1,183	1,076
Ethiopia	208	221	228	240	306	319	343	613	1,049
Senegal	74	392	466	539	686	686	760	870	980
Haiti	608	672	711	784	735	613	613	735	907
Nepal	-	-	-	-	761	980	735	869	900
Bhutan	502	564	623	637	1,414	1,422	735	882	895
Jordan	245	137	639	681	647	521	978	1,082	891
Swaziland	1,225	1,225	1,103	1,103	858	700	882	980	882
Côte d'Ivoire	196	368	515	662	445	686	717	818	858
Occupied Palestinian Territories	-	-	-	-	-	2,370	1,880	924	858
Kenya	233	294	355	466	613	637	637	637	686
Lao People's Democratic Republic	417	392	441	564	515	490	711	686	686
Central African Republic	270	282	306	328	382	490	588	515	539
Puerto Rico	720	823	765	725	716	427	647	460	478

Table 9. Continued.

Country	1970/71	1975/76	1980/81	1985/86	1990/91	1995/96	2000/01	2005/06	2008/09
Sudan	833	760	282	355	245	368	424	441	441
Mali	196	221	245	257	306	306	306	353	392
Bangladesh	299	197	210	226	213	196	221	245	382
Mozambique	417	490	539	490	637	319	325	343	343
Togo	191	218	270	282	294	297	297	299	331
Suriname	241	233	215	234	329	368	241	314	330
Azerbaijan	-	-	-	-	-	539	515	473	309
Benin	270	294	294	294	294	294	294	294	306
Malasya	478	306	306	201	265	260	294	294	294
Afghanistan	449	461	368	314	279	284	284	233	284
New Zeland	43	70	158	221	294	515	123	172	216
Somalia	159	179	191	206	218	196	196	221	213
Dominica	44	48	58	67	78	106	176	172	176
Liberia	127	145	162	176	172	172	172	172	176
Guinea-Bissau	-	-	15	34	108	123	123	123	147
Guyana	206	267	261	279	157	101	196	123	147
Sri Lanka	195	171	295	146	82	69	130	145	141
Republic of Trinidad and Tobago	437	294	164	167	184	365	95	129	129
Montenegro	-	-	-	-	-	-	-	-	128
Albania	81	147	265	324	250	93	64	127	125
Philippines	273	268	472	455	226	203	197	142	119
Zambia	49	64	76	83	96	86	86	88	86
Congo	59	69	74	71	69	61	61	56	56
Guadalupe	10	9	3	13	8	4	22	54	54
Saint Vincent and the Grenadines	5	6	7	6	22	24	24	39	42
Malta	-	-	-	-	-	-	29	34	39
Tajikistan	-	-	-	-	-	-	25	15	29
Tonga	49	56	61	66	25	12	17	25	28
Grenada	21	19	23	21	22	21	22	22	22
United Republic of Tanzania	-	-	-	-	-	-	15	22	22
Martinique	9	12	20	16	13	13	17	20	20
Reunion Island	-	1	4	9	22	48	29	18	18
Fiji	10	20	5	12	13	12	17	17	17
Botswana	5	7	11	12	12	12	15	15	15
France	56	64	34	63	45	28	13	16	15

Table 9. Continued.

Country	1970/71	1975/76	1980/81	1985/86	1990/91	1995/96	2000/01	2005/06	2008/09
French Guiana	3	5	5	4	4	14	15	15	15
French Polynesia	6	7	5	1	3	5	4	9	15
East Timor	-	-	-	-	-	17	7	15	15
Santa Lucia	4	4	5	7	8	16	35	13	14
Croatia	-	-	-	-	-	19	13	15	14
Burkina Faso	-	-	5	12	12	12	12	13	13
Brunei Darussalam	12	9	6	6	6	5	6	8	8
Russian Federation	-	-	-	-	-	-	-	9	5
Bosnia and Herzegovina	-	-	-	-	-	2	2	3	3
Cook Islands	123	93	31	20	11	10	4	2	2
Guam	1	0	1	1	1	1	1	2	2
Seychelles	0	0	0	1	1	1	1	1	1
Kuwait	1	1	1	0	0	2	0	0	0
Montserrat	0	0	0	0	0	0	0	0	0
Djibouti	-	-	-	-	0	0	0	0	0
Serbia and Montenegro	-	-	-	-	-	46	83	100	-
USSR	3,353	3,775	3,676	3,529	7,157	-	-	-	-
Yugoslav SFR	28	69	114	54	286	-	-	-	-

10. Evolution of Brazilian orange production

Orange farming occurs in all Brazilian States. With more than 800 thousand hectares, the orange is the most planted fruit in Brazil. Comparatively speaking, orange groves occupy an area 20 times larger than that for apple orchards, 10 times larger than that for mango plantations and for vineyards and almost twice the amount of land destined for banana plantations. The orange groves are situated outside the State of São Paulo, which boasts 70% of the planted area.

The planted area for citrus farming in Bahia and Sergipe has almost doubled in size since the beginning of the 1990s, when these two States represented 7% of the orange groves in Brazil and today, with this increase, they already boast 13% of the national area. In this same period, the area in the State of Paraná grew fourfold, in Alagoas it is now 7 times larger and in other States, such as Goiás, Pará, Amapá and Acre they have doubled the size of their plantations. Production in the States less specialized in citrus farming is mostly destined for the domestic market for fresh fruit, to meet the growing demand due to the greater purchasing power of the Brazilian population. The capacity of these new citrus belts in the Northeast and South of Brazil to increasingly meet the demand for fresh oranges consumed by the population in the North, Northeast and Mid-West of this country contributed in the last decade to the reduction of fresh fruit in Sao Paulo and Triangulo Mineiro region, which went from between 80 million and 100 million boxes to between 30 million and 40 million boxes per year.

Although there has been growth in the area of orange plantation in these regions, the total planted area in Brazil has decreased by around 8% since the beginning of the 1990s. This shrinkage was not accompanied by a reduction in the quantity of boxes picked. On the contrary, there was an increase of 22%. This inversion is the result of an impressive increase in productivity. The national average of 380 boxes per hectare in 1990, jumped this year to 475 boxes per hectare. If citriculture were the same today as it was 20 years ago, a further 280 thousand hectares, approximately, would be necessary to reach current production levels.

11. Specialty of the major producing countries

The world's foremost citrus producers have different destinations for their production, forming a mix between industrial processing (production of juice), domestic consumption of fresh fruit and export of fresh fruit. Brazil, the world's largest producer of oranges, sends 70% of its production for industrial processing, with São Paulo and the Triângulo Mineiro region sending 86% of their production to this market. To get an idea of the importance of the processing industry for these two Brazilian regions, you must realize that, in Brazil, no other fruit is produced for industrial purposes on a level similar to the production of oranges.

The United States are very similar to Brazil in terms of the destination for their production. Around 78% of its fruit is bound for processing, since production in Florida is almost exclusively focused on juice, reaching a level of 96%. Mexico concentrates on producing fresh fruit, resulting in a small amount of orange juice, around 60 thousand tons per year, of which 50 thousand tons are exported to the United States and ten thousand to Europe. China focuses on the supply of fruit for fresh consumption, which is the destination of 93%

of production, but has received private and government investments that are set to drive juice production in the next few years. Spain is renowned for its exports of fresh fruit with high aggregated value, since it produces seedless fruit with an excellent color and external appearance, albeit very acidic and with a low juice content (Graph 10).

The number of boxes of oranges needed to produce one ton of FCOJ at 66° Brix is a determining factor in establishing the vocation of a citrus farming region. This indicator, called industrial yield, is what determines the attractiveness of the production from this region for the manufacturer of orange juice.

In the 2009/10 harvest, the best industrial yield was attained in Florida, where 226 boxes of oranges were needed to produce one ton of FCOJ. In the Brazilian citrus belt, located in São Paulo and the Triângulo Mineiro region, 257 boxes were needed. In the States of Bahia, Sergipe, Paraná and Rio Grande do Sul 264 boxes were needed. In China the number was 324 boxes, in the Mediterranean Zone it was 328 boxes and in Arizona, California and Texas it was 353 boxes. The poorest yield was from Turkey, with 481 boxes per ton of juice.

The fruit with the highest juice content, a characteristic born of several factors, the majority of which are not controlled by humans, is what has maintained São Paulo and Florida as leaders in the production of orange juice for decades, while the other citriculture centers have had to specialize in the production and packing of fresh oranges. This fact also means that

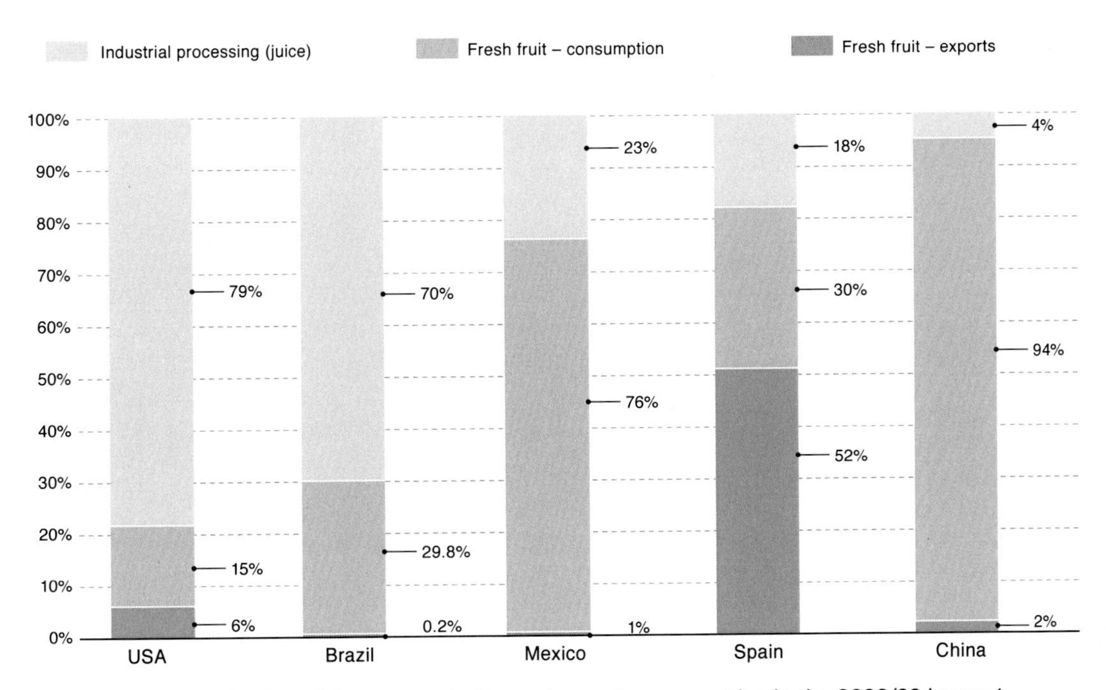

Graph 10. **Destination of the orange in the main producer countries in the 2008/09 harvest.**
Source: prepared by Markestrat, based on data from the Spanish Department of Agriculture, FAO, USDA and IBGE.

a box of oranges produced in Florida has an additional gain of 14% in the content of orange juice compared with a similar box of oranges produced in São Paulo.

This higher yield was important gave the oranges produced in Florida an unrivalled competitive advantage. Other factors that increase American competitiveness are the proximity of the industrial citriculture region in Florida to the American consumer, direct access of the Florida producers to low cost credit on the financial market (making it unnecessary for the industry to make pre-payments on working capital funds for the harvest),the absence of exchange rates for the currency, the non-application of import taxes on local production (around U$ 415 per ton of FCOJ) and the result of decades of investment in marketing to convince the American consumer that orange juice "produced 100% in Florida" is a better quality product, which would justify the higher prices.

12. Orange juice production

In the last 15 harvests, from 1995/96 to 2009/10, the drop in world production of orange juice has been 13% (equivalent to 308 thousand tons), with the largest reductions occurring in Florida (295 thousand tons and in the citrus belt of São Paulo and the Triângulo Mineiro region (31 thousand tons (Graph 11). Although they have diminished, these regions still lead world production of orange juice, with 81% of all production.

In the States of Paraná, Bahia, Sergipe, Rio Grande do Sul, Pará, Goiás and Rio de Janeiro the market for fresh fruit uses up 77% of the production from these regions. Nevertheless, depending on the rises in the price for juice, these Brazilian States, as in the case of other

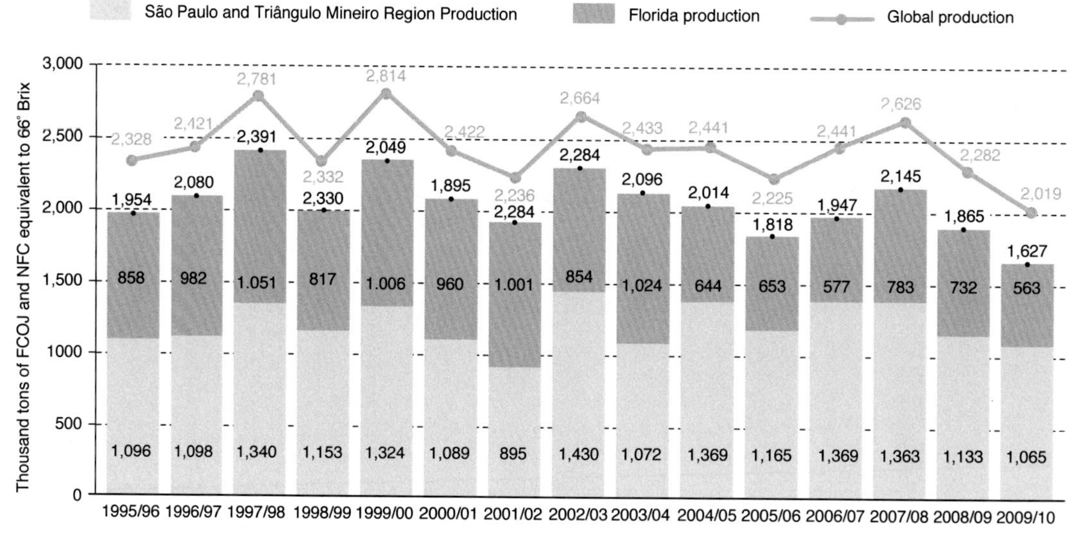

Graph 11. Evolution of the world's production of orange juice.

Source: prepared by Markestrat based on CitrusBR.

countries of lesser importance in the production of juice, such as South Africa, China, Spain, Greece, Italy, India, Mexico, Pakistan and Turkey, among other countries, switch to processing more oranges, and together they produce 150 thousand more tons of concentrated juice than normal (Graph 12). The increase in the supply of juice from these regions tends to push prices down.

However, the greatest influence on the price in terms of supply comes from oscillations in production, stocks and availability of juice from Brazil, the United States and the Mediterranean Region where, together, they account for 96% of the orange juice produced in the world. This being so, São Paulo, despite representing half of the global production, suffers from the impact of these other countries and regions, whose joint production is just as significant as the production from the State of São Paulo, making it difficult for Brazilian companies.

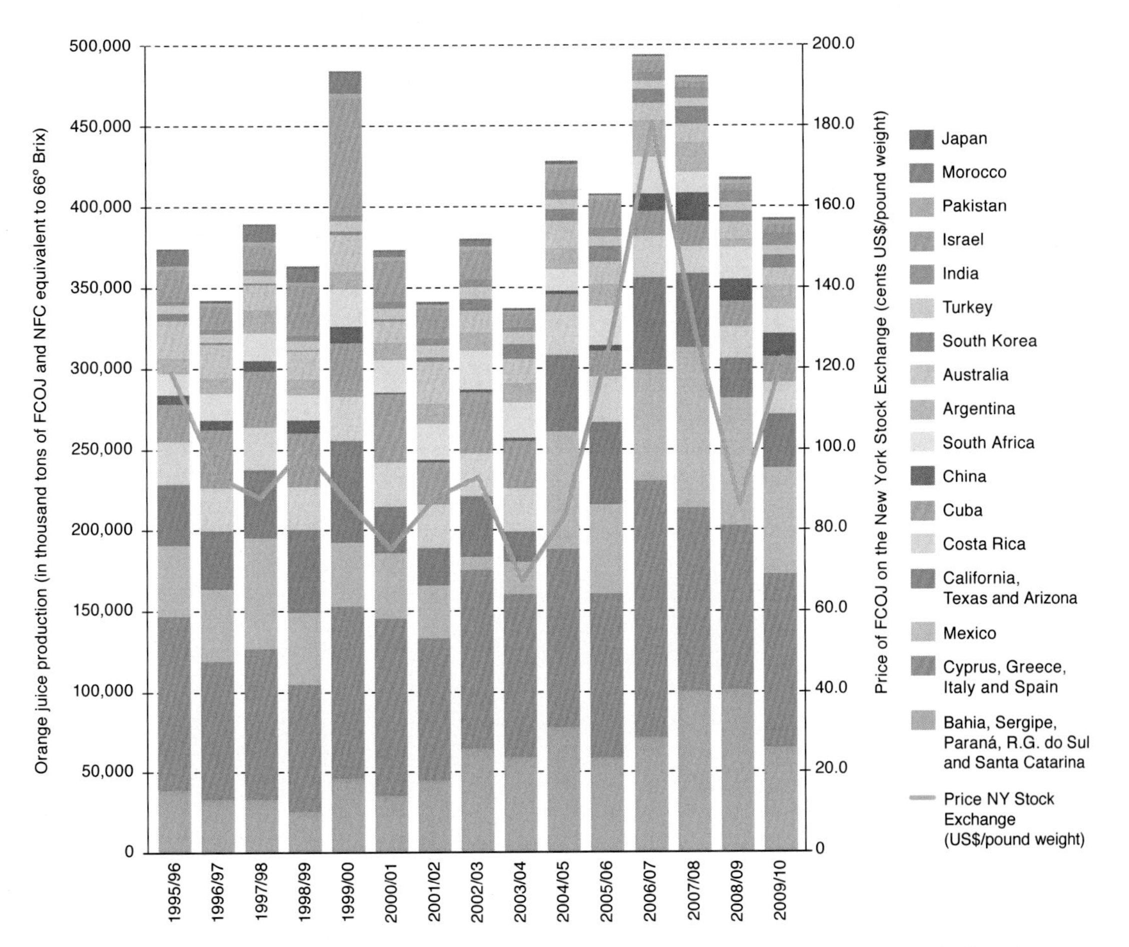

Graph 12. World production of orange juice in other farming regions except São Paulo/Triângulo Mineiro and Florida.

Source: prepared by Markestrat, based on data gathered from CitrusBR.

Orange juice from Brazil is known for its high quality. Brazil is the number one producer and exporter, boasting 53% of the world's orange production (Table 10 and 11) and exporting approximately 98% of this production. The type of juice produced is dictated by consumer behavior in markets with more spending power, which in recent years have come to favor NFC over FCOJ, since it is a product with a more pleasant flavor, closer to that of freshly squeezed juice and because it has a more healthy image. The first production of NFC in Brazil began in 1999/2000 on an experimental basis. In 2000 the first exports took place, but it was only in 2002/03 that NFC was registered by the Secex (Export Dept.), separately from the exports of FCOJ.

To meet the growing consumer demand, the production of FCOJ gradually gave way to the production of NFC. From 2003 to 2009, the industry invested around US$ 900 million

Table 10. Quantity of boxes of oranges (40.8 kg) destined for world production of juice.

Country/Region	1995/96	1996/97	1997/98	1998/99	1999/00
São Paulo and Triângulo Mineiro Region	271,120	271,420	322,740	270,690	308,900
Florida	195,277	219,701	235,632	177,359	224,289
Cyprus, Greece, Italy and Spain	37,132	37,475	41,838	35,417	44,730
Mexico	11,029	11,029	17,157	11,103	10,049
Bahia, Sergipe, Paraná, Rio Grande do Sul and Santa Catarina	9,960	8,195	8,324	6,433	11,589
California, Texas and Arizona	11,341	8,804	10,937	13,425	18,579
Costa Rica	6,500	6,500	6,500	6,500	6,500
Cuba	5,613	8,652	8,456	8,088	8,088
China	2,108	2,328	2,451	3,162	3,971
South Africa	4,289	5,392	5,588	7,696	7,549
Argentina	3,162	3,186	4,804	3,186	3,431
Australia	4,779	7,353	4,657	5,613	7,402
South Korea	1,644	307	360	131	561
Turkey	2,059	2,181	1,814	2,377	2,696
India	782	1,001	1,257	1,154	1,199
Israel	4,657	5,049	4,044	1,593	3,725
Pakistan	672	686	699	639	667
Morocco	2,353	147	2,892	2,083	3,186
Japan	49	25	49	49	49
Total	574,526	599,432	680,198	556,698	667,161

Source: prepared by Markestrat based on CitrusBR.

in the production, storage and international distribution of NFC. It is estimated that for each box processed, stored and transported in the form of this juice to overseas customers, the investment required is three times higher than that for FCOJ. The investments made for this new product at 11.5th Brix made it possible for exports of NFC to jump from 278,572 tons in 2003 to 939,442 tons in 2009, the equivalent of approximately 171 thousand tons of FCOJ at 66° Brix or at 13% of the total amount of juice exported by Brazil. These investments in the production of NFC are likely to be amortized over a period of 10 to 15 years, according to the specialists consulted.

2000/01	2001/02	2002/03	2003/04	2004/05	2005/06	2006/07	2007/08	2008/09	2009/10
266,450	212,500	324,220	242,070	329,900	265,330	316,550	317,650	287,790	274,120
213,635	221,843	194,579	233,790	142,836	142,091	122,519	165,906	154,814	127,436
44,387	36,618	42,917	40,539	43,701	47,549	54,118	37,206	39,093	34,804
9,804	8,333	1,961	4,902	18,137	13,725	17,157	24,510	19,608	15,931
9,014	11,461	16,479	16,222	20,659	15,966	19,082	26,883	26,522	17,375
9,134	6,490	10,470	5,083	13,463	13,547	16,403	10,540	7,588	11,093
6,578	6,500	6,500	6,500	6,500	6,800	6,225	4,044	4,804	5,025
10,196	6,373	9,118	6,985	2,623	3,799	3,799	3,799	3,799	3,799
368	564	515	613	686	1,029	3,554	6,005	4,461	4,951
6,961	7,672	5,956	5,564	3,088	6,961	6,324	4,902	6,740	6,740
3,505	3,922	3,676	3,922	4,167	4,412	7,353	5,833	1,740	4,681
4,167	8,309	4,167	4,657	5,392	4,338	3,333	3,676	3,309	3,456
464	795	1,922	2,353	1,971	2,458	2,147	2,721	1,738	2,157
2,623	3,064	3,064	3,064	2,451	2,574	2,574	2,451	2,451	2,451
1,311	1,262	1,407	942	1,600	1,625	1,685	2,092	2,155	2,221
2,083	1,569	1,324	686	1,863	1,495	2,059	1,029	1,324	1,618
651	628	583	604	667	844	844	844	844	844
931	441	1,078	123	147	147	147	147	147	147
49	49	49	49	49	49	49	49	49	49
592,312	538,391	629,984	578,666	599,899	534,737	585,920	620,288	568,976	518,898

Table 11. World production of orange juice (all types of juice in tons at 66° Brix).

Country/Region	1995/96	1996/97	1997/98	1998/99	1999/00
São Paulo and Triângulo Mineiro Region	1,095,780	1,098,000	1,339,970	1,152,860	1,324,150
Florida	858,032	981,790	1,051,448	816,520	1,005,905
Cyprus, Greece, Italy and Spain	107,339	86,331	93,234	79,083	106,856
Mexico	44,318	44,811	68,939	44,614	40,379
Bahia, Sergipe, Paraná, Rio Grande do Sul and Santa Catarina	38,672	31,821	32,320	24,978	45,000
California, Texas and Arizona	37,552	36,137	41,724	50,493	62,798
Costa Rica	26,916	26,916	26,916	26,916	26,916
Cuba	23,242	35,827	35,015	33,492	33,492
China	5,524	6,102	6,423	8,286	10,405
South Africa	13,271	16,683	17,290	15,511	23,145
Argentina	9,848	9,848	15,068	9,848	10,606
Australia	22,571	19,992	14,485	17,352	22,790
South Korea	4,956	1,232	1,447	526	2,252
Turkey	4,880	5,158	4,289	5,622	6,375
India	2,607	3,335	4,190	3,847	3,998
Israel	19,697	14,513	14,550	30,530	72,879
Pakistan	2,241	2,288	2,330	2,129	2,222
Morocco	10,051	559	10,732	9,356	13,591
Japan	148	98	197	197	197
Total	2,327,643	2,421,442	2,780,565	2,332,160	2,813,957

Source: prepared by Markestrat based on CitrusBR.

13. Brazil's citrus belt (São Paulo and Triângulo Mineiro)

Brazilian citriculture has undergone changes in technological standards that are even more notable in São Paulo and the Triângulo Mineiro region, the so-called citrus belt, the source of more than 80% of the oranges produced in this country. Despite the fact that this is a single geographical area, there are significant differences between the citriculture from one region to another. For didactic purposes, to aid understanding of the particularities of the regions, in this work the citrus belt has been divided into five producing regions denominated (1) Northwest, (2) North, (3) Center, (4) South, according to their geographical positions within the State of São Paulo, and (5) Castelo, a name derived from its position in relation to the Castelo Branco freeway. Figure 3 shows these regions, also indicating the location of the processing plants.

2000/01	2001/02	2002/03	2003/04	2004/05	2005/06	2006/07	2007/08	2008/09	2009/10
1,089,010	894,520	1,429,660	1,072,450	1,369,260	1,164,500	1,369,210	1,362,720	1,132,850	1,064,650
959,726	1,000,672	854,355	1,024,009	644,461	653,301	577,349	782,504	731,799	562,663
109,811	87,344	110,598	100,947	110,303	100,553	158,626	111,983	100,756	106,173
39,886	33,485	7,879	19,795	72,977	55,644	68,939	100,455	78,788	66,477
35,000	44,500	63,986	58,302	77,003	59,275	70,875	100,058	101,433	65,775
29,419	22,913	37,622	19,448	47,099	50,705	56,929	45,260	24,257	31,385
27,240	26,916	26,916	26,916	26,916	28,158	25,779	16,746	19,892	20,806
42,220	26,388	37,755	28,925	10,860	15,731	15,731	15,731	15,731	15,731
963	1,477	1,477	1,773	1,970	2,955	10,833	17,727	13,788	15,265
20,475	22,681	24,198	22,583	13,898	25,173	22,868	13,295	19,712	14,687
10,833	12,121	11,364	12,121	12,879	13,636	22,727	18,030	5,379	14,470
12,829	25,682	12,879	14,394	16,667	13,409	10,303	11,363	10,228	10,686
1,866	3,194	7,724	9,455	7,919	9,875	8,627	10,935	6,986	8,667
6,201	7,244	7,244	7,244	5,795	5,795	5,274	5,042	5,100	5,100
4,370	4,207	4,690	3,140	5,332	5,415	5,617	6,972	7,184	7,403
25,606	18,712	18,712	8,864	14,773	17,629	7,445	3,644	4,826	5,811
2,170	2,093	1,944	2,013	2,223	2,812	2,812	2,812	2,812	2,812
3,939	1,576	4,333	492	591	591	591	591	591	591
197	197	197	197	197	197	197	197	197	197
2,421,763	2,235,922	2,663,534	2,433,067	2,441,122	2,225,354	2,440,732	2,626,066	2,282,307	2,019,348

Thanks to this change in technological standards, production in the belt has grown significantly, reaching 317.4 million boxes in the 2009/10 harvest, an increase of 16% over the course of the decade (Graph 13). Among the changes that have taken place in citriculture, special emphasis is placed on the density of trees per hectare. In 1980, the most widely used density of plantation was 250 trees per hectare, rising to 357 trees per hectare in the 1990s, then to 476 trees per hectare at the beginning of 2000 and, currently, most modern groves consist of 833 trees per hectare.

In addition to the density, other important factors that led to this increase in productivity were the use of better quality saplings, grown in screened enclosures and with genetic lineage; better combinations of rootstock and varieties better suited to each type of climate and soil; enhancement of the knowledge applied to improve the management of the groves and the quality of phytosanitary control; in addition, there has been an intensification and increase in the use of irrigation in the regions with the greatest shortages of water, contributing to achieving the current level of 130 thousand hectares of irrigated orange groves in São Paulo and the Triângulo Mineiro region.

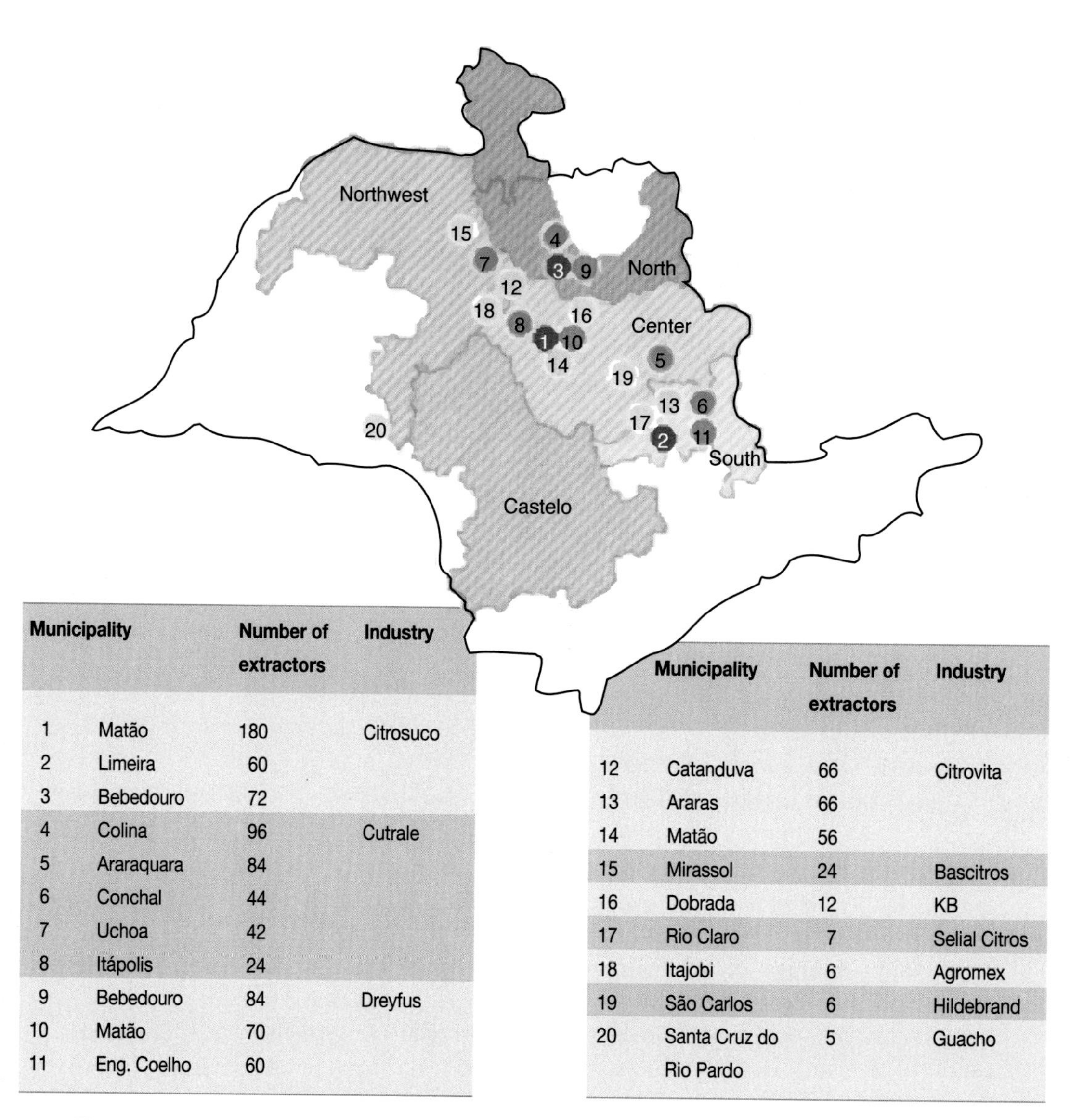

	Municipality	Number of extractors	Industry
1	Matão	180	Citrosuco
2	Limeira	60	
3	Bebedouro	72	
4	Colina	96	Cutrale
5	Araraquara	84	
6	Conchal	44	
7	Uchoa	42	
8	Itápolis	24	
9	Bebedouro	84	Dreyfus
10	Matão	70	
11	Eng. Coelho	60	

	Municipality	Number of extractors	Industry
12	Catanduva	66	Citrovita
13	Araras	66	
14	Matão	56	
15	Mirassol	24	Bascitros
16	Dobrada	12	KB
17	Rio Claro	7	Selial Citros
18	Itajobi	6	Agromex
19	São Carlos	6	Hildebrand
20	Santa Cruz do Rio Pardo	5	Guacho

Figure 3. Division of the regions in the citrus belt.

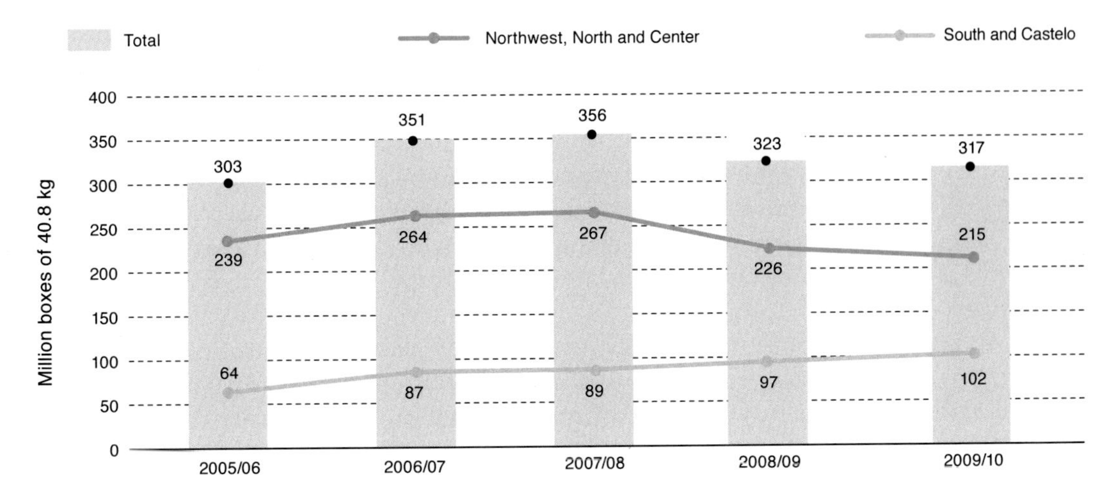

Graph 13. Orange production in the citrus belt.
Source: prepared by Markestrat based on CitrusBR.

Another important point is determining the optimum time for renewing the grove in accordance with the levels of productivity per hectare. In considering the longevity of the groves, a period of 20 years is normally used as a reference for the lifespan of a tree with economically viable productivity in order to amortize investment, but observation of the industrial groves reveals that this age can be surpassed, reaching 26 years in industrial groves in the Bebedouro region, 24 years in the Mogi-Guaçu region and 22 years in Araraquara.

In the citrus belt, there has also been a migration of citriculture from the regions in the North, Northwest and Center to the South and Castelo regions, where the microclimate is more favorable to citrus farming. This movement, which intensified at the start of 2000, was initially motivated not only by the climatic conditions, but also by the better distribution of rain over the course of the year and also by the lower price for land and as an option to reduce the rate of sudden death in citrus trees and CVC, which on this new frontier does not represent a threat to the groves and the control of which is onerous and depends on high technology.

Nowadays, the main factors determining the transfer of citriculture to new areas are mitigation of the risk of greening, which spread to 239 municipal districts in the State of São Paulo, almost half of the citrus farming towns, as well as the massive expansion of São Paulo's sugarcane plantations into areas that were previously planted with orange groves, resulting in low-productivity and insufficient profitability. Areas with less infestation or where greening has not been encountered are now much more desirable and, this being so, in the coming decades the groves are likely to move further and further away from where the factories are located nowadays. The demand for land in the southern regions of the State has led to an even greater increase in value than that witnessed in land in the north.

The most outstanding region in the citrus belt has undoubtedly been Castelo. Between 2005 and 2009, there was an increase of 89% in the total of trees in this region, taking it from last place to second in the number of trees. Also, 42% of the new trees (from zero to two years) in the citrus belt are in this region, indicating the growing importance of this region in the coming years for participation in the total production of the belt (Figure 4).

Due to the greater fluctuations in temperature in the South and Castelo regions, the color of the fruit tends to be better, in comparison with the fruit produced in the other regions. However, since the average temperature in these regions is lower, the fruit is usually more acidic and has a lower Brix than the oranges from the Center, North and Northwest. To obtain the specifications for the juice required by the consumer market, it is necessary to make a blend using the juices from both regions.

Orange production in the citrus belt, as well as the destination of the same, has changed over time, as it has become evident that there was an increase in the supply for industry and, consequently, a reduction in the supply for fresh consumption. The production destined for industry went from 76% of the total production total for the citrus belt in 1995 to 86% of the

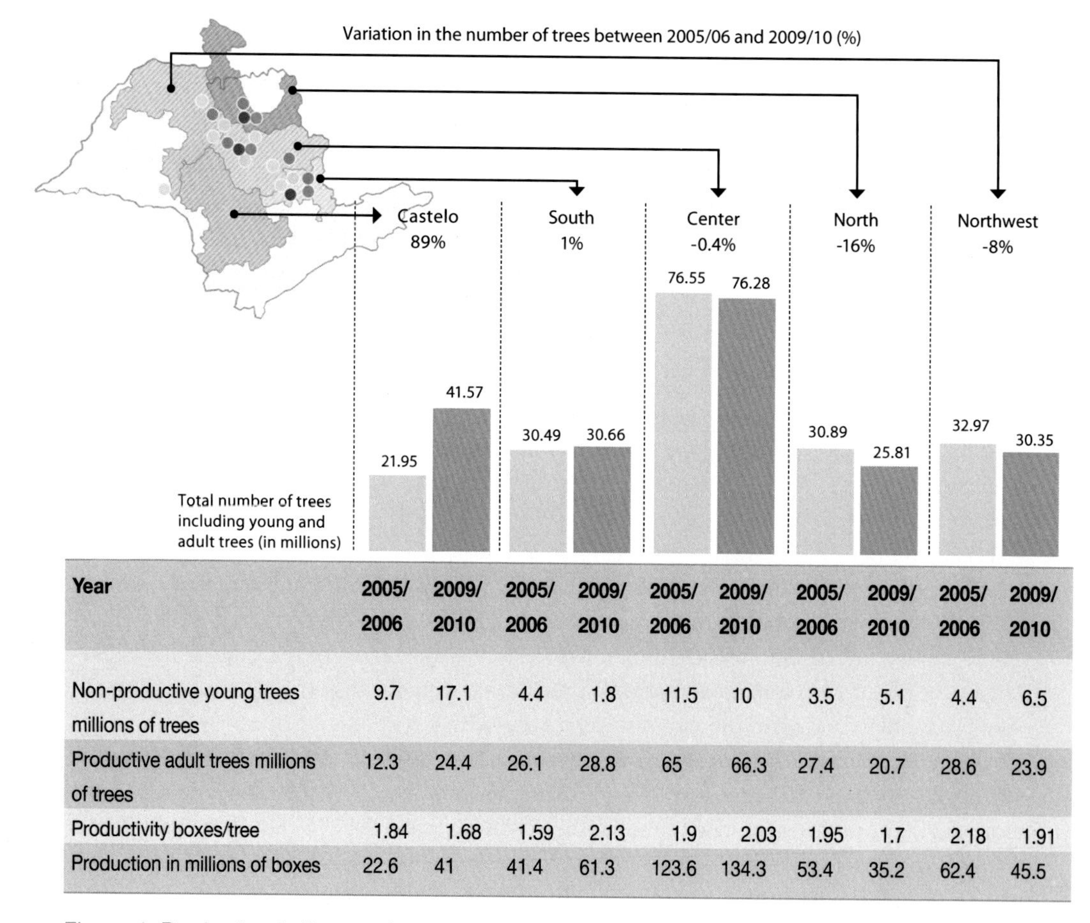

Year	2005/2006	2009/2010	2005/2006	2009/2010	2005/2006	2009/2010	2005/2006	2009/2010	2005/2006	2009/2010
Non-productive young trees millions of trees	9.7	17.1	4.4	1.8	11.5	10	3.5	5.1	4.4	6.5
Productive adult trees millions of trees	12.3	24.4	26.1	28.8	65	66.3	27.4	20.7	28.6	23.9
Productivity boxes/tree	1.84	1.68	1.59	2.13	1.9	2.03	1.95	1.7	2.18	1.91
Production in millions of boxes	22.6	41	41.4	61.3	123.6	134.3	53.4	35.2	62.4	45.5

Figure 4. Production indicators for the citrus belt regions.

Source: prepared by Markestrat based on CitrusBR

production in 2009, in other words, a growth of 10%. On the other hand, it can be seen that the fruit destined for fresh consumption, which represented 24% in 1995 and dropped to 14% in 2009, suffered a reduction of 10% (Graph 14).

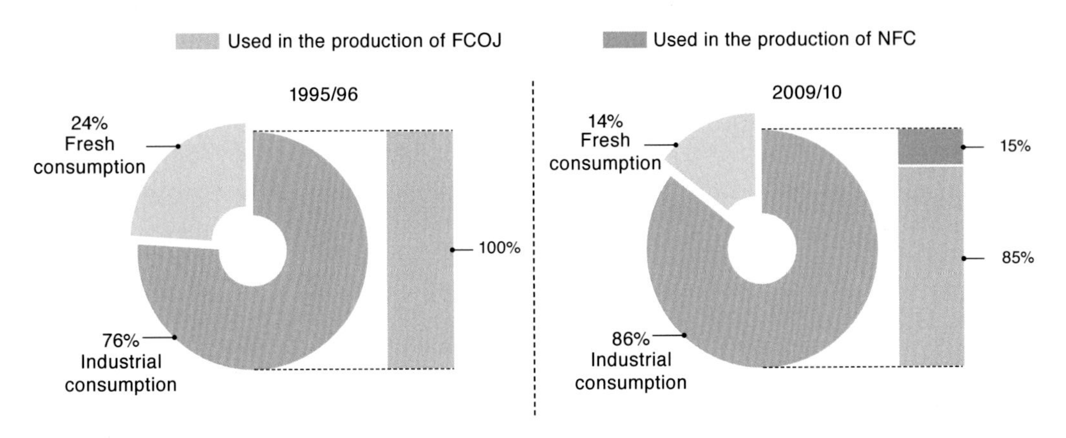

Graph 14. Destination of the orange production from the citrus belt in the 1995/96 and 2009/10 harvests.

Source: prepared by Markestrat, based on CitrusBr.

14. Comparison between the production in São Paulo/ Triângulo Mineiro region and Florida

The States of São Paulo, in Brazil, and Florida, in the United States, dominate the world supply of orange juice with 81% of the total. This high level of concentration in two producer regions is a rare event in the ambit of agricultural commodities, but the strength of these two regions used to be greater in the past. In the 1990s, the total joint orange production for the two regions was around 600 million boxes, and in the first decade of 2000, production reached levels around 500 million boxes. This means 100 million fewer boxes of oranges were produced. Despite the fact that production has diminished in both regions, the largest decrease was in Florida.

In the 2003/04 harvest, the North American State produced the equivalent of 87% of the orange production from São Paulo, but in 2009/10 it only attained half this amount. This reduction intensified in the wake of the hurricanes that devastated the region in 2004 and 2005, spreading citrus canker and greening, diseases for which the solution is eradication. This fact, added to the problems of restriction and more expensive labor costs, contamination of the water tables, higher prices for land and other issues related to climatic risks such as droughts, hurricanes and harsh winters have been quashing the enthusiasm of American citrus farmers.

This is reflected in the drop in the number of trees in recent years. Since 2004/05, there has been a reduction of approximately 19% in the number of trees (loss of 15 million). In addition to these factors, there has been no renewal of the groves. Currently, around 45% of the trees are more than ten years old and the number of young trees, aged two years at most, is no more than 10%. This increase in the average age of the groves is reflected in the productivity of the trees, which has also been diminishing (Graph 15).

Part of the reduction in these areas is due to the rising price of land, which has led many producers from both regions to abandon this activity. In Florida, the real estate boom has taken the place of thousands of productive trees in the areas close to the towns, where there is now

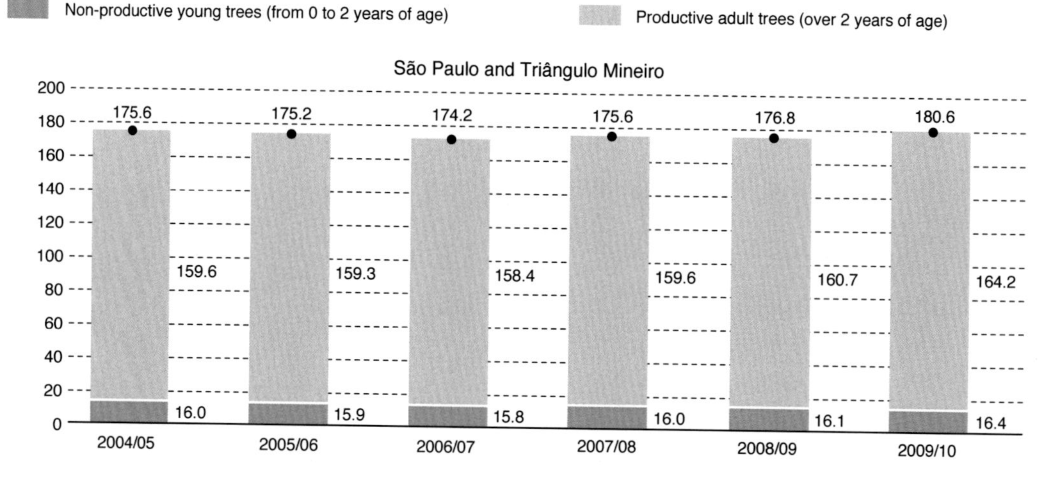

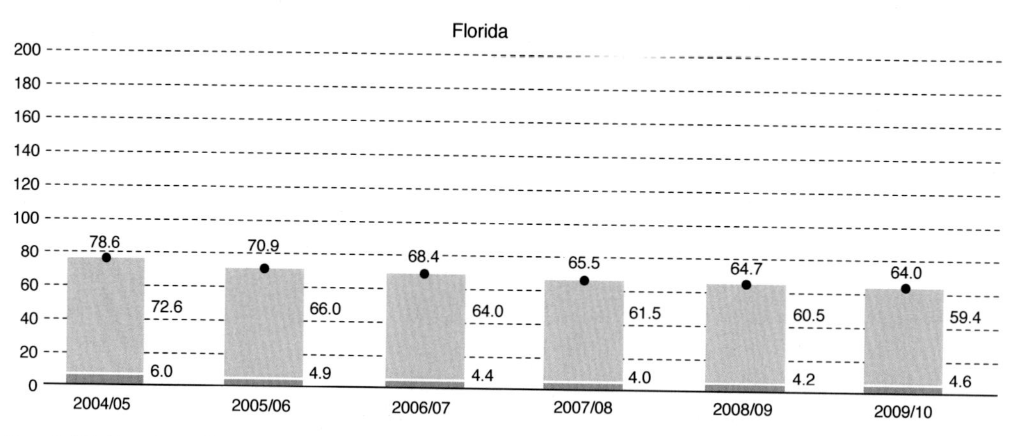

Graph 15. Number of non-productive young trees (from zero to two years) and productive adult trees (over two years) in the São Paulo and Triângulo Mineiro regions and in the state of Florida.

Source: prepared by Markestrat based on CitrusBR and USDA.

a growing number of residential condominiums, especially in the period from 2003 to 2007, when there was a strong rise in real estate prices (Table 12).

Production in São Paulo suffered a sharp drop in the 2003/04 harvest, but since then it has been recovering. However, problems with greening, excessive rain and Colletotrichum are likely to result in reduced production from the 2010/11 harvest, which will put even more distance between the levels of productivity for the Brazilian citrus belt and those witnessed in Florida. While in the last decade Florida obtained an average productivity of 2.56 boxes/tree, the Citrus Belt covering São Paulo and the Triângulo Mineiro region obtained an average of 2.06 boxes/tree in the same period. This being so, the average productivity in the Citrus Belt covering São Paulo and the Triângulo Mineiro region is 25% lower than that of its foremost competitor, the State of Florida (Table 13).

In recent years there has also been a migration in citrus farming to the southern regions of the States, both in Florida and in São Paulo. In São Paulo, the oranges produced in the South of the State present a lower industrial yield in relation to the other producer regions. Nevertheless, the citriculture located further south in the State presented significant gains in agricultural productivity, indicating that in the future the industrial yield in boxes per ton of FCOJ will tend to be lower than that generally seen in the past.

Table 12. Evolution of land prices (US$/hectare) in municipal districts within the Brazilian citrus belt and in Florida.

Municipal districts within the belt	2001	2003	2005	2007	2009	Variation in period from 2001 to 2009
Araraquara	2,352	2,699	4,936	9,919	9,126	288%
Bauru	2,940	2,851	4,923	10,153	9,742	231%
Campinas	4,621	4,595	7,312	10,534	10,140	119%
Itapetininga	2,016	2,532	3,772	5,117	5,478	172%
Piracicaba	3,781	3,424	4,971	10,328	10,393	175%
Pirassununga	2,520	3,342	5,929	10,666	9,029	258%
Ribeirão Preto	4,032	5,159	8,049	10,647	9,633	139%
São José do Rio Preto	2,352	3,147	5,869	8,831	9,247	293%
Southern Florida	15,827	14,647	24,583	39,810	29,842	89%

Source: prepared by Markestrat based on data from Agrianual 2010 and the 'Florida Land Value Survey' IFAS various surveys.

Table 13. Details of citriculture in the Brazil's citrus belt and in Florida.

Harvest	Productive adult trees (in millions)			Agriculture yield (40.8 kg boxes/tree)		Orange production (millions of 40.8 kg boxes)		
	São Paulo and Triângulo	Florida	Total	São Paulo and Triângulo	Florida	São Paulo and Triângulo	Florida	Total
1988/89	97.5	36.8	134.3	2.20	3.99	214.0	146.6	360.6
1989/90	102.2	40.7	142.9	3.05	2.71	311.2	110.2	421.4
1990/91	108.7	44.1	152.8	2.27	3.44	246.8	151.6	398.4
1991/92	115.1	49.6	164.6	2.22	2.82	256.0	139.8	395.8
1992/93	125.3	56.6	181.9	2.54	3.30	318.1	186.6	504.7
1993/94	146.0	61.7	207.7	2.07	2.83	302.2	174.4	476.6
1994/95	156.8	69.3	226.1	1.96	2.97	307.3	205.5	512.8
1995/96	162.8	75.3	238.1	2.19	2.70	356.3	203.3	559.6
1996/97	172.6	78.5	251.1	2.15	2.88	371.0	226.2	597.2
1997/98	179.9	78.6	258.5	2.38	3.10	428.2	244.0	672.2
1998/99	171.5	79.6	251.1	1.97	2.34	338.5	186.0	524.5
1999/00	166.0	78.7	244.7	2.63	2.96	436.0	233.0	669.0
2000/01	162.5	79.6	242.1	2.15	2.81	349.7	223.3	573.0
2001/02	162.3	77.6	239.8	1.68	2.96	272.8	230.0	502.8
2002/03	158.8	78.0	236.9	2.31	2.60	367.5	203.0	570.5
2003/04	157.8	75.4	233.2	1.77	3.21	278.6	242.0	520.6
2004/05	159.6	72.6	232.1	2.37	2.06	377.8	149.8	527.6
2005/06	159.3	66.0	225.3	1.90	2.24	303.4	147.7	451.1
2006/07	158.4	64.0	222.4	2.22	2.02	351.0	129.0	480.0
2007/08	159.6	61.5	221.1	2.23	2.77	356.0	170.2	526.2
2008/09	160.7	60.5	221.2	2.01	2.68	323.3	162.4	485.7
2009/10	164.2	59.4	223.6	1.93	2.25	317.4	133.6	451.0

Source: prepared by Markestrat based on CitrusBR and USDA.

Oranges destined for fresh consumption (millions of 40.8 kg boxes)			Oranges destined for industrial consumption (millions of 40.8 kg boxes)			Industrial yield (40.8 kg boxes/ton of juice at 66° Brix)		Total orange juice production (thousand tons 66° Brix)		
São Paulo and Triângulo	**Florida**	**Total**	**São Paulo and Triângulo**	**Florida**	**Total**	**São Paulo and Triângulo**	**Florida**	**São Paulo and Triângulo**	**Florida**	**Total**
34.8	8.5	43.2	179.3	138.1	317.4	261	220	687.8	628.0	1,315.8
50.0	5.9	55.9	261.2	104.3	365.5	259	220	1,007.6	474.1	1,481.8
47.3	12.5	59.7	199.5	139.1	338.6	242	220	823.5	632.7	1,456.2
43.8	11.6	55.3	212.2	128.2	340.4	236	220	898.0	583.1	1,481.1
43.1	10.7	53.9	275.0	175.9	450.9	257	220	1,070.0	799.7	1,869.7
53.8	9.9	63.7	248.4	164.5	412.9	237	220	1,048.1	748.3	1,796.4
62.9	10.4	73.3	244.4	195.1	439.5	233	229	1,047.7	853.2	1,900.9
85.2	8.0	93.2	271.1	195.3	466.4	247	228	1,095.8	858.0	1,953.8
99.6	6.5	106.0	271.4	219.7	491.1	247	224	1,098.0	981.8	2,079.8
105.4	8.4	113.8	322.7	235.6	558.4	241	224	1,340.0	1,051.4	2,391.4
67.8	8.6	76.4	270.7	177.4	448.0	235	217	1,152.9	816.5	1,969.4
127.1	8.7	135.8	308.9	224.3	533.2	233	223	1,324.2	1,005.9	2,330.1
83.3	9.7	92.9	266.5	213.6	480.1	245	223	1,089.0	959.7	2,048.7
60.3	8.2	68.5	212.5	221.8	434.3	238	222	894.5	1,000.7	1,895.2
43.3	8.4	51.7	324.2	194.6	518.8	227	228	1,429.7	854.4	2,284.0
36.5	8.2	44.7	242.1	233.8	475.9	226	228	1,072.5	1,024.0	2,096.5
47.9	7.0	54.8	329.9	142.8	472.7	241	222	1,369.3	644.5	2,013.7
38.1	5.6	43.7	265.3	142.1	407.4	228	217	1,164.5	653.3	1,817.8
34.4	6.5	40.9	316.6	122.5	439.1	231	212	1,369.2	577.3	1,946.6
38.3	4.3	42.6	317.7	165.9	483.6	233	212	1,362.7	782.5	2,145.2
35.5	7.6	43.1	287.8	154.8	442.6	254	212	1,132.9	731.8	1,864.6
43.3	6.2	49.4	274.1	127.4	401.6	257	226	1,064.7	562.7	1,627.3

15. Stratification of production by producer profile in Brazil's citrus belt

In citriculture, as in any economic activity faced with tight margins, there is an urgent need to increase productivity in order to reduce the production cost for each box of oranges. This need becomes even more pressing for those citrus farmers that send their produce to industry, where the prices tend to be lower than in the market for fresh fruit. In order to generate profits from the sale of oranges for industrial processing, it is necessary to have large-scale production. It also requires compliance with labor and environmental legislation, such as the registering of employees, respect for the use of approved pesticides and the waiting periods between their application and the harvest of the fruit, as well as appropriate disposal of containers. Such factors are essential to international juice buyers.

These requisites are more easily met by the larger enterprises, which use high technology and are generally an ideal size for adequate sizing of the equipment, as well as higher purchasing power to buy supplies. However, 87% of the producers in the citrus belt are of a smaller size (11,011 producers) farming properties with less than 20 thousand trees. This collection of producers represents just 21% of the trees in the citrus belt. The other producers are divided as follows: 12% are medium-sized producers (between 20 and 199 thousand trees), consisting of 1,496 producers that have 32% of the trees in the belt, and 1% are large-sized producers (more than 200 thousand trees), with a total of 120 producers owning 47% of the trees in the citrus belt (Table 14).

This data is from CitrusBR, which for the first time traced the profile of producers in the São Paulo and Triângulo Mineiro citrus belt based on the register of the citrus farmers that supplied oranges to the industry in the 2009/10 harvest. This was done using the registers held by the industry with regard to the citrus farmers, containing all the registrations required for the survey, in other words, area size, number of trees and volume produced. The registers used represent around 80% of all fruit processed by the industry. These registers were provided by the industries, on an individual basis, to one of the world's top international independent auditing companies, which compiled the data with absolute secrecy and supplied the average values to the Association, in the same way that other associations in Brazil gather data relating to their members.

In this survey, it was seen that the share of larger properties in the number of trees in the citrus belt has increased. This fact indicates that producing oranges using high technology and a large production scale is an economically viable activity. In 2001, properties with more than 400,000 trees comprised 16% of the total number of orange trees in Brazil's citrus belt, while in 2009 this percentage jumped to 39%. Properties with 10,000 to 199,000 trees comprised 61% of the trees in the citrus belt and this figure fell to 40% by 2009. A similar scenario occurred with soybeans in the state of Mato Grosso, where agricultural groups doubled the size of their croplands over the past five years; around 20% of all soybeans grown in Mato Grosso were grown by the 20 largest groups, according to data from Reuters. Concentration in agriculture is a reality in almost all crops and is being studied internationally. Only strong associations,

Table 14. Stratification of growers in the agricultural belt, by number of trees.

Trees × 1000	2001			2006			2009		
	Trees (%)	Growers (%)	Number of growers	Trees (%)	Growers (%)	Number of growers	Trees (%)	Growers (%)	Number of growers
>400	16.15	0.15	23	33.65	0.35	46	39.25	0.4	51
200 to 399	7.65	0.25	38	8.05	0.55	73	7.35	0.55	69
100 to 199	10.6	0.7	105	8.1	1.05	139	8.95	1.3	164
50 to 99	12.4	1.75	263	11.45	2.7	356	10.75	2.95	372
30 to 49	12.3	3.15	473	7.7	3.35	442	7	3.5	442
20 to 29	8.95	3.9	585	5.5	3.8	502	5.3	4.1	518
10 to 19	16.45	14.5	2,175	9.45	11.35	1,498	8	11.15	1,408
<10	15.45	75.55	11,333	16.15	76.9	10,151	13.4	76.05	9,603
Total	100%	100%	15,000	100%	100%	13,200	100%	100%	12,627

Source: prepared by Markestrat based on information from CitrusBR, considering data obtained from member-organizations.

cooperatives and modern models of integrated grower networks have been efficient and speedy in reducing this concentration.

In 2009, 44% of the planted acreage in the citrus belt had productivity below the minimum levels to earn income. These croplands produce 280 boxes or oranges per hectare, on average. This is a big difference compared to other properties that account for the remaining 56% of the total number of hectares, which produce an average of 909 boxes per hectare (Table 15).

This dynamic that is occurring in the citrus sector explains why less efficient producers – unable to compete with the more efficient ones – have left the orange-growing business and started to focus on other crops. Those who have remained in citrus farming must find a more appropriate path for each farm profile, i.e. define a particular strategy for running their orange groves, such as cost leadership, differentiation, or diversification (Figure 5).

The strategy of "cost leadership" focuses on ongoing efforts to keep down production and distribution costs, requiring strong skills in production processes, operating yields, and acquisition of lower-priced inputs. It is this low-cost strategy that should be followed by orange-growers whose produce is intended for the juice industry, because profit margins are lower in this sector. In order to be competitive when adopting this strategy, production of scale is required. Acquiring inputs at lower prices, for example, depends on high purchasing volume and/or planning to make purchases during periods in which there is lower demand for the input. If inputs are purchased at times when demand is high, they must be bought in bulk in order to get discounts from the vendors.

However, those who choose the "differentiation" strategy to conduct their business will produce oranges with attributes that are highly valued by consumers of fresh fruit, who

Table 15. Stratification by productivity range (boxes per hectare) of orange production – 2009/10 growing season.

Range of productivity (boxes/ha)	% of hectares	% of boxes	Volume of boxes produced per productivity range (× million)	Yield (boxes/ha)
> 1,400	2%	5%	16	1,655
1,100 - 1,399	7%	13%	41	1,209
800 - 1,099	19%	29%	92	933
500 - 799	28%	30%	95	639
200 - 499	36%	21%	67	345
< 200	8%	2%	6	138
Total	100%	100%	317.4	607
Total > 500	56%	77%	244.4	909
Total < 499	44%	23%	73	280

Source: prepared by Markestrat based on information from CitrusBR, considering data obtained from member-organizations.

are willing to pay higher prices for the differentiated product. The differentiation strategy has been achieved by producing a higher-quality fruit during times when supply is low. It is a fact that smaller properties have advantages in adopting the differentiation strategy; they are able to carry out more intense monitoring of factors such as pruning, nutrition management, irrigation and application of growth regulators. This strategy has proved attractive to small and medium-sized orange growers who sell their produce as fresh orange.

Another strategy adopted by small and medium-sized growers in the citrus belt, following in the footsteps of citrus growers in the South of Brazil, is "diversification." These growers in the citrus belt can grow similar crops, such as guava, passion fruit, mango and grapes, among others, reducing the risk of concentrating on one single activity, always taking care not to deviate too much from the primary focus of the business.

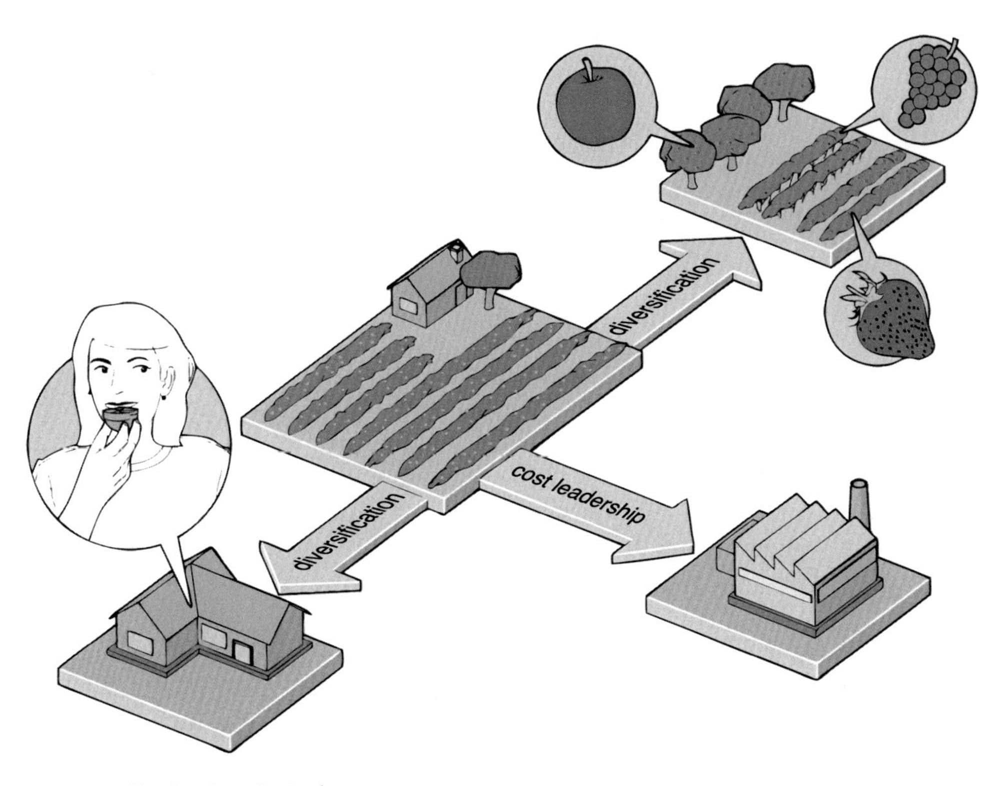

Figure 5. Production strategies.

16. Varieties in the orchards

Diversification of varieties in the orchards is important because it distributes the harvest throughout the year, avoiding the concentration of supply in a few months, and allowing the citrus farmers to sell during times of higher prices and still allow industry to extend the period of processing oranges for juice production.

Today, the orange orchards in the state of São Paulo contain 55% late-season varieties of trees (Natal and Valencia), 23% early-season varieties (Hamlin, Westin, Ruby and Pineapple), and 22% mid-season varieties (Pera Rio). Orange growers' preference for late-season varieties, due to higher productivity, occurred at the expense of mid-season varieties, which are well accepted in the fresh-fruit market, leading to a shortage in the supply of oranges particularly in the month of September, thus causing greater competition between the juice industry and fresh-fruit market during this period. In addition, being well accepted on the fresh-fruit market, the Pera Rio variety has a higher content of soluble solids (juice). These two factors, plus the production shortage at the very time in which Pera Rio oranges are being harvested, allow this variety to earn higher prices than other varieties destined for the juice industry (Graph 16).

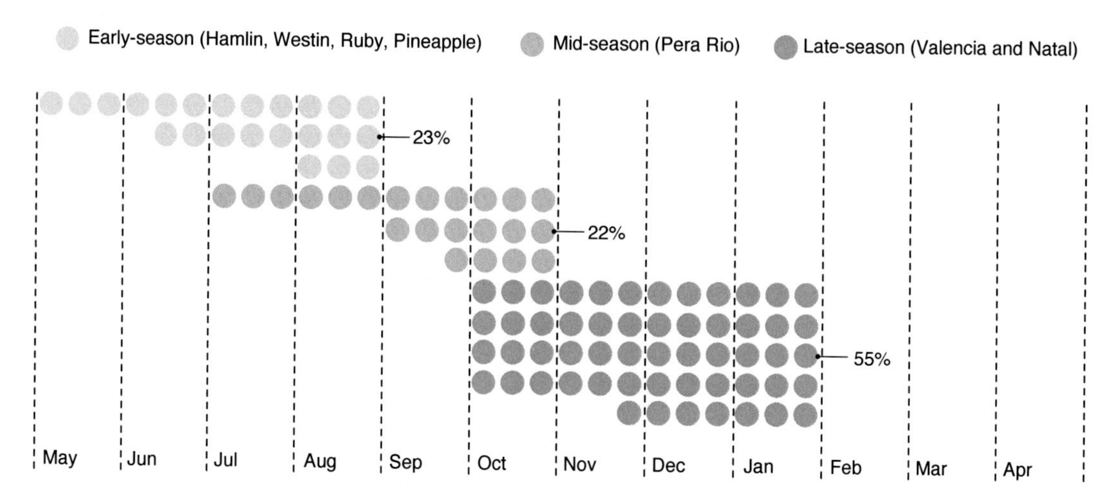

Graph 16. **Harvest period by variety and percentage of production.**
Source: prepared by Markestrat based on CitrusBR.

In order to reduce the period of supply shortage, growers are changing the profile of their orchards, increasing early-season varieties and decreasing late-season varieties. In orchards with trees aged zero to two years, early-season varieties represent 29% of the total number of trees. However, numbers of mid-season varieties still tend to be low.

Planting different varieties is also a way to manage the control of diseases and reduce the impact of bad weather. Enhancement of citrus varieties has been done with traditional improvement techniques. More recently, investments in research and development have focused on the study of genetically modified varieties, with the aim of combating diseases that are economically significant for the sector. This is the case in the research conducted by the Centro de Citricultura Apta Citros, in pursuit of orange varieties tolerant of citrus canker and/or resistant to greening. However, these studies might not be widely used in Brazil, since the main market for Brazilian production – the European market – is still uncompromising when it comes to genetically modified products.

17. Pests and diseases in the Brazil's citrus belt

This is undoubtedly a major threat to the Brazilian citrus industry. During the last decade, four diseases were responsible for the eradication of 39 million citrus trees in the citrus belt of São Paulo and Triângulo Mineiro. As a consequence, the average annual mortality rate, which previously hovered around 4.5% per year, jumped to 7.3%. Adopting an average yield of two boxes of oranges per tree, it is estimated that citrus canker, CVC, sudden death, and citrus greening accounted for an annual reduction of around 78 million boxes, which – when compared with 317 million boxes harvested in 2009/10 – represents a decrease in harvest of roughly 20% (Table 16).

Citrus canker is a bacterial disease that causes the premature falling of leaves and fruit, and reached its apex in the 1990s. It is the oldest of the four major diseases present in Brazil. CVC (citrus variegated chlorosis) -a bacterial disease that affects the vascular system of the trees, reducing the size of the fruit to that of a golf ball – is the disease that has caused the most damage to date, originating in the northern and northwestern regions of São Paulo state and later migrating to the center of the citrus belt. Sudden death – a vascular disease that can kill the tree in 12 months – developed mainly in the northern regions of São Paulo state and in the Triângulo Mineiro region, in orange trees grafted onto the rootstock of the Rangpur lime (Citrus Limonia Osbeck). Finally, there is greening, the most recent bacterial disease and the one that causes greatest concern among citrus growers due to the speed with which it has spread, from its point of origin in the central region of São Paulo to other regions.

Table 16. Trees eradicated in the São Paulo and Triângulo Mineiro citrus belt due to the 4 major diseases that affect citriculture (in thousands of trees).

	2000	2001	2002	2003	2004	2005	2006	2007	2008	2009	Total
Canker	795	191	71	164	177	153	186	151	115	240	2,243
CVC	678	2,406	2,380	1,023	2,887	4,043	3,320	3,299	3,276	3,070	26,382
Greening	-	-		-	-			5,330			5,330
Sudden death	-	-			5,158			-	-	-	5,158

Source: prepared by Markestrat based on the annual percentage of eradication of trees released by Fundecitrus, weighted by the number of trees in the citrus belt reported by CitrusBR.

18. Impact of climate change on citrus growing

In recent years, several reputable organizations have warned about the risks of climate change to world agriculture. Things are no different in the citrus sector. Data from the National Institute of Meteorology (Inmet) show that there has been a gradual increase in average temperature in several Brazilian states. The comparison is obvious when comparing the averages between two 30-year periods, i.e. from 1930 to 1960, in relation to the numbers obtained in the measurements from 1960 to 1990. Although the curve has remained the same, one can clearly see that the state of São Paulo is warmer. In some regions, such as Limeira and São Jose do Rio Preto, the numbers obtained between 1995 and 2009 indicate that the temperature is, on average, about two degrees Celsius higher than the historical average (Figure 6).

The result has been the worsening of weather conditions, as the studies suggest, including scarcer rainfall in the northern part of São Paulo state and more concentrated rainfall in the southern region, according to a report released by the São Paulo State Department of Water. Crossing such information with what has been happening in recent years in the São Paulo citrus industry, we can see why the need for irrigation in orchards in the northern part of the state is

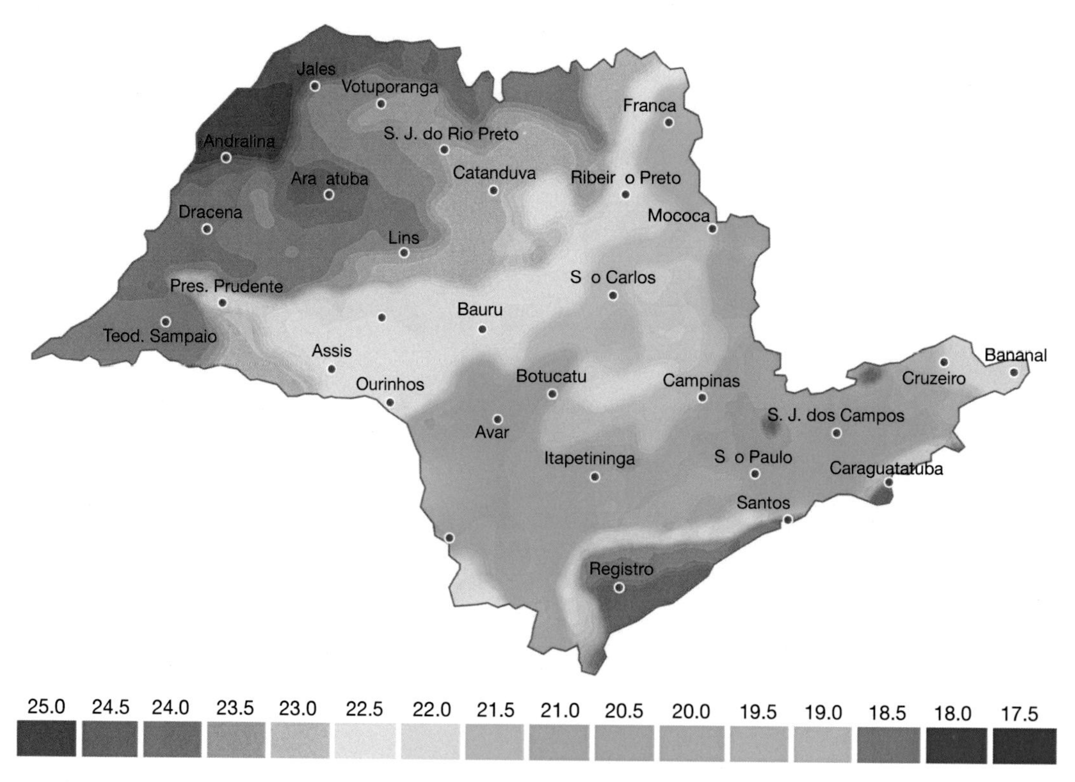

Figure 6. Annual normal average temperature of the state of Sao Paulo (1961-1990).
Source: IAC.

so pressing today, as opposed to two to three decades ago. Another point observed – this time by the Intergovernmental Panel on Climate Change (IPCC) – indicates a worsening of extreme weather phenomena. In 2010, for example, under the effects of La Nina, some regions had more than 100 days of drought, a far-from-trivial fact when compared to historical averages.

Therefore, the production of oranges in the northern part of the citrus belt requires other techniques that were unnecessary in the 1960s and 1970s, such as use of rootstocks other than Rangpur lime and irrigation. These techniques require not only more structured citrus-growing practices, but also much higher investments than those of 40 years ago. The result of this in the coming years – if the effects of climate change continue to grow – may accentuate the change in the geography of citrus farming in the state of São Paulo, shifting even more strongly to the southern regions of the state. The map shows the temperature gradient within the state of São Paulo for the period from 1961 to 1990. One can clearly see the difference in temperature between the regions further south compared to the northern regions of the state.

19. Cost of orange production

Production cost is an important issue because it is an excellent planning tool and can be used by growers to decide on investments in one crop or another, and assist in management and decision-making as to whether or not to remain in the activity. Perennial crops have higher production costs, but also generally the highest returns.

This was the conclusion drawn from the comparative analysis of operating costs of coffee, sugarcane and soybean production, using data from Agrianual/AgraFNP for the years 2005 to 2009. Compared with the cost of orange growing, the operating costs for producing these crops were lower, with the exception of 2006 (Graph 17).

However, the average yield of oranges in this period was second only to coffee, which was 193% higher, showing the relative attractiveness of oranges vis-à-vis sugarcane and soybeans, whose profitability were 41% and 81% lower (respectively) in relation to oranges, although it's a crop that requires a high degree of specialization (Graph 18).

Besides being a relevant issue, production cost is also a controversial topic, due to the wide range of factors, which involve a set of activities and input of management techniques, varying considerably regarding quantity, frequency and efficiency depending on the technology adopted, soil and climatic characteristics, pressure from pests and diseases, and sanitary requirements of each location. In addition to the foregoing aspects, the productivity achieved can have a major impact on production cost because – despite the fact that harvesting and shipping costs are variable – other expenses and costs of cultivation are almost entirely fixed per tree and per hectare. Therefore, the higher the productivity of one tree or one hectare, the lower the production cost of oranges on the tree. The consequence of this high number of variables is the discrepancy in production costs found in the orange sector (Graph 19).

From 2002/03 to 2009/10, the average operating cost of 100% of the oranges produced by industries in each of the seasons was determined for the first time. The data was also compiled confidentially and individually by one of the largest international independent audit firms.

19. Cost of orange production

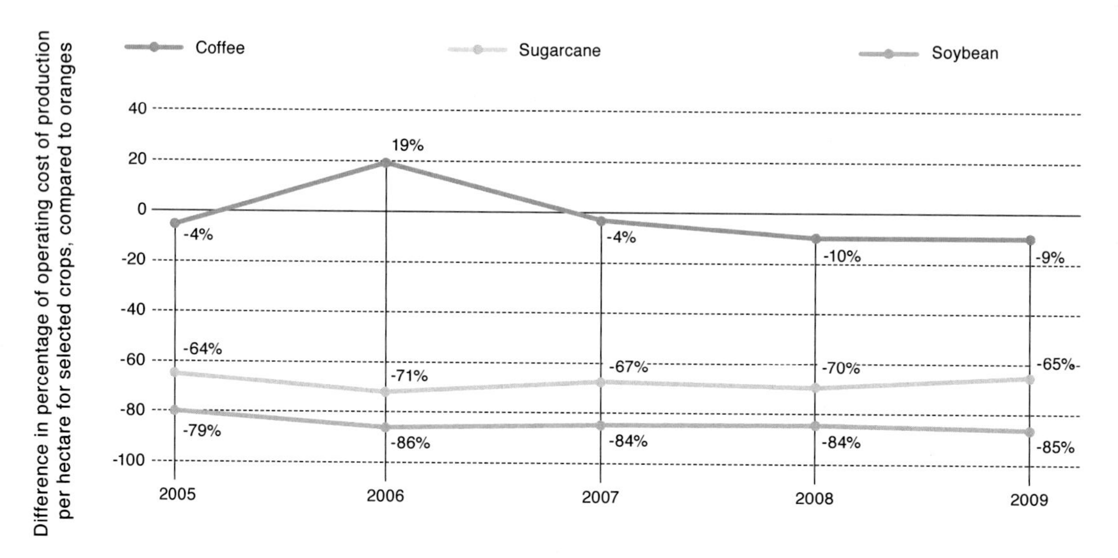

Graph 17. **Difference in percentage of the operating cost of production per hectare for selected crops, compared to oranges.**

Cost of production of orange, soybean, coffee and sugarcane crops, published in Agrianual/Agra FNP – Average values between 2005 and 2009.

Source: prepared by Markestrat based on Agrianual/AgraFNP.

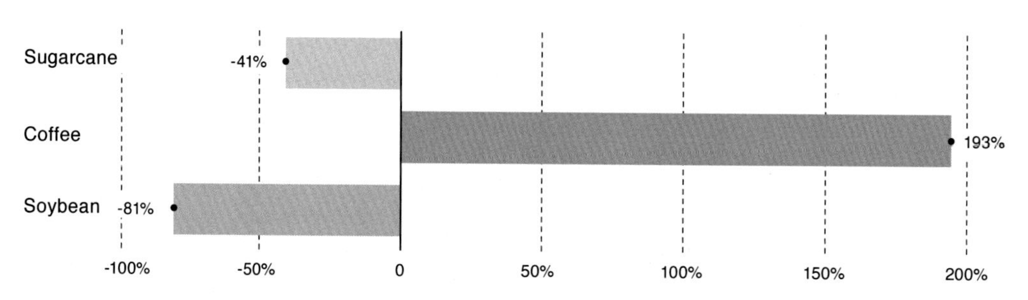

Graph 18. **Difference in percentage of average financial result of selected crops in relation to the financial results for oranges.**

Financial results for orange, soybean, coffee and sugarcane crops, published in Agrianual/Agra FNP – Average values between 2005 and 2009.

Source: prepared by Markestrat based on Agrianual/AgraFNP.

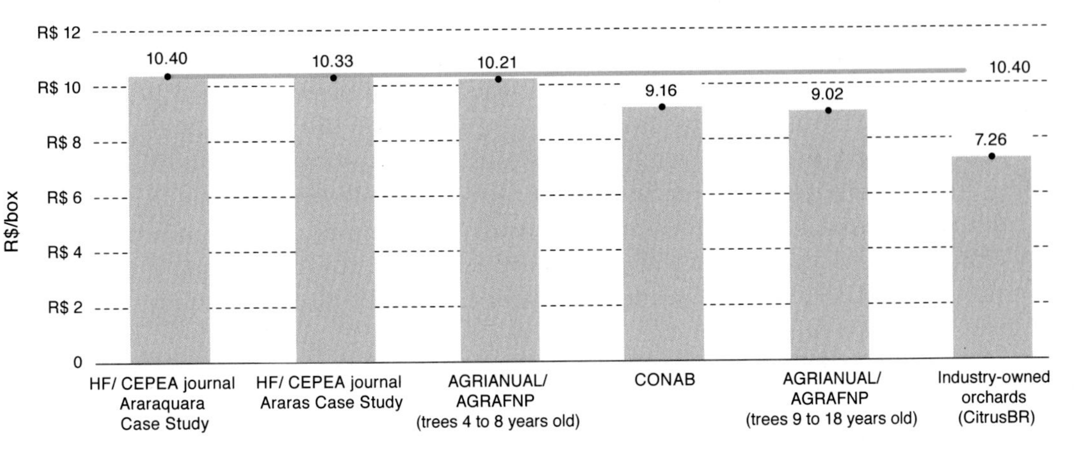

Graph 19. **Comparison of operating costs of orange production per box, based on different sources for 2009/10 growing season.**

Operating cost of production of oranges per 40.8 kg box, including production on the tree + harvest + shipping + ITR without depreciation or financial expenses and without remuneration of capital.

Source: prepared by Markestrat based on Agrianual/AgraENP, Conab CitrusBR, and CEPEA

Agrianual/AgraFNP Cost in 2009:

Density: 408 tress/hectare

Araraquara region

Module 100 hectares

As the purpose is to compare the operating costs, costs related to maintenance and depreciation of improvements were excluded.

CONAB 2009/10 Cost

Density: 400 tress/hectare

Bebedouro Region

As the purpose is to compare operating costs, financial expenses, cost of depreciation, and income of factors were excluded.

HF/CEPEA Journal – Case Study in Araras for the 2009/10 growing season, prepared by Larissa Pagliuca, Mayra Viana, Margarete Boteon, Keila Inoue, Fernanda Geraldini, and João Paulo Bernardas Deleo.

Araras region.

Density: 434 trees/hectare.

Module 128 hectares

100% dryland farming.

Cost of harvesting and shipping of the portion destined for industry were considered.

As the purpose is to compare operating costs, costs related to the working capital, CARP and the opportunity cost of the land were excluded.

HF/CEPEA Journal – Case Study in Araraquara for the 2009/10 growing season, prepared by Larissa Paglluca, Mayra Viana, Margarete Boteon, Keila Inoue, Fernanda Geraldini, and João Paulo Bernardas Deleo.

Araraquara Region.

Density: 324 trees/hectare.

Module 214 hectares

79% of the area irrigated with simple drip line. Cost of harvesting and shipping of the portion destined for industry were considered.

As the purpose is to compare operating costs, costs related to the working capital, CARP and the opportunity cost of the land were excluded.

These figures represent the operating cost for production of about 35% of the oranges processed by juice industries, originating from company-owned orchards scattered throughout the Citrus Belt, from Itapetininga in southern São Paulo state to Uberlândia in the Triângulo Mineiro region (Table 17).

Since this is strictly a matter of operating cost, the following are excluded: cost of establishing orchards between zero and three years (CAPEX and financing), land lease costs, depreciation and amortization of machinery and equipment, depreciation costs of gains in valuation of land, Fundecitrus fees, financing expenses for working capital for the harvest, and financial income or expenses.

Based on the analysis of this operational cost of production over the past eight years, it is clear why there has been so much talk in the industry about monitoring production costs and the search for alternatives to reduce such costs in virtue of the rising cost of inputs. The price of diesel fuel, for example, increased by over 100% in Brazilian Reais between 2002 and 2009, raising expenditure on mechanized faming activities and shipping. In this same period, the cost of harvesting rose approximately 160% (including amounts paid in wages, compulsory and optional payroll charges, NR 31 compliance measures, and PPE). In 2009, harvesting was twice as expensive as shipping. However, historically, these were costs of a similar magnitude. Between the 2002/03 and 2009/10 growing seasons, costs for harvest and freight rose from 35% to 44% of the operating costs of orange production (Graph 20).

In all, the rise in operational cost of orange production was around 202% in dollars between the 2002/03 and 2009/10 growing seasons, from US$ 1.31/ box to US$ 3.96/box. Highlighted during this period was the 2008/09 season, in which there was a significant increase of approximately 25% over the previous crop year, driven mainly by the rising price of fertilizers.

In 2002/03, the orange groves owned by Brazilian juice industries in São Paulo and Triângulo Mineiro had more competitive production costs than the groves in Florida (3.3 times lower), but this advantage in favor of Brazilian industry was reduced in 2008/09 (Graph 21 and Table 18).

The higher costs of orange production are evidence of the need to rethink the management of citrus enterprises, to adopt production planning that involves the determination of long-term objectives and targets, to establish actions and allocate resources to achieve them. It is also incumbent upon government agencies to implement integrated support mechanisms in this rethinking of the productive activity, due to this sector's importance in generating jobs and income.

Table 17. Average operating cost of orange production in orchards owned by juice industries (40.8 kg box) audited by an international firm.

Items		Harvest							
		2002/03	2003/04	2004/05	2005/06	2006/07	2007/08	2008/09	2009/10
A. Labour (wages, compulsory and optional payroll charges, PPE, outsourced manpower)	US$	0.27	0.39	0.36	0.51	0.52	0.65	0.80	0.91
B. Pesticides and herbicides	US$	0.22	0.34	0.25	0.36	0.34	0.38	0.46	0.49
C. Fertilizers (fertilizers and soil correctives)	US$	0.16	0.27	0.23	0.29	0.28	0.32	0.41	0.41
D. Electricity	US$	0.02	0.03	0.02	0.03	0.05	0.05	0.07	0.06
E. Expenditure with its own vehicles and outsourced services	US$	0.07	0.11	0.09	0.11	0.13	0.16	0.20	0.17
F. Maintenance, conservation and other general expenses	US$	0.11	0.16	0.14	0.17	0.19	0.25	0.25	0.17
G. Total expenditures on the tree = A + B + C + D + E + F	US$	0.85	1.30	1.10	1.49	1.51	1.82	2.18	2.21
H. Harvest (Wages, compulsory/ optional payroll charges, NR 31, PPE)	US$	0.26	0.35	0.37	0.54	0.65	0.81	0.97	1.20
I. Shipping costs (internal removal, freight charges to the factories and road tolls)	US$	0.20	0.25	0.29	0.37	0.41	0.48	0.54	0.56
J. Total expenditures ex factory, in dollars = G + H + I	US$	1.31	1.90	1.75	2.41	2.58	3.15	3.67	3.96
Average exchange rate of disbursement during harvest time (reais/dollar)	R$	3.23	2.98	2.84	2.37	2.17	1.87	1.97	1.83
J. Total expenditures ex factory, in Brazilian reais = G + H + I	R$	4.25	5.65	4.98	5.67	5.59	5.81	7.28	7.26
Average distance travelled by oranges from company-owned orchards to the factory		161 km	162 km	137 km	133 km	125 km	126 km	133 km	132 km

Source: CitrusBR.

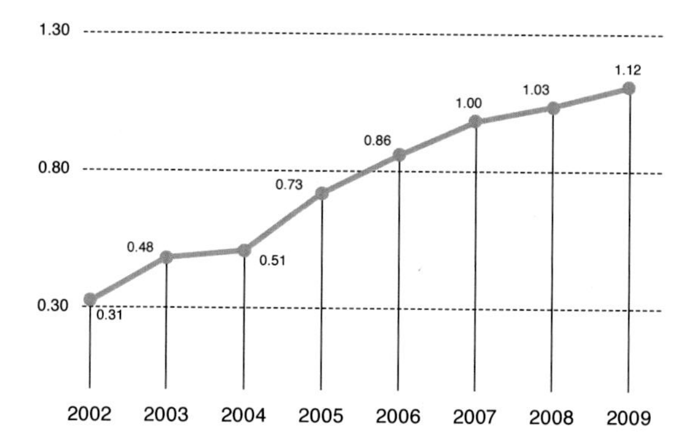

Cost of harvesting oranges (US$/box)

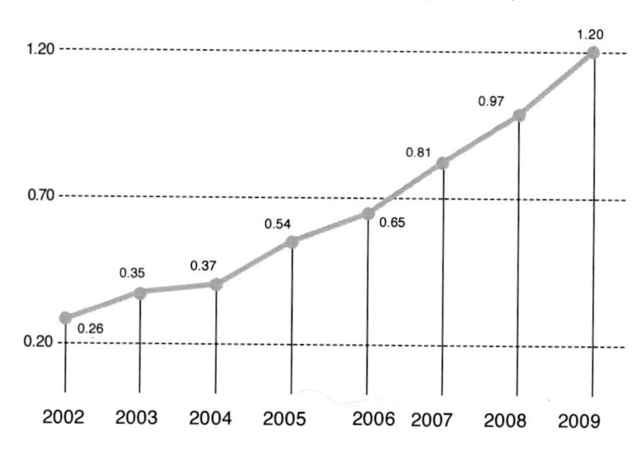

Shipping cost (US$/box)

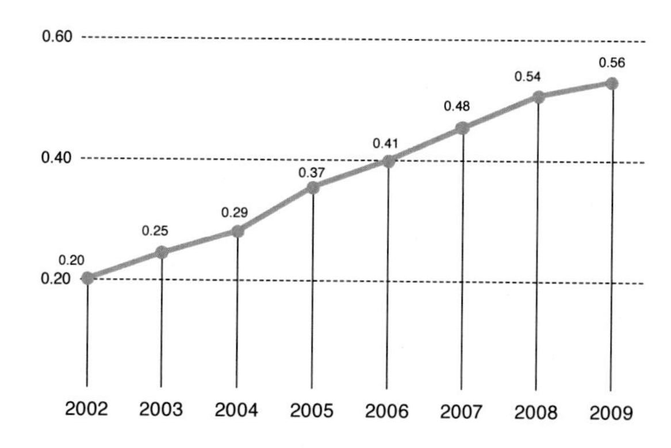

Graph 20. Price of diesel, shipping cost and the harvest cost in São Paulo for high-tech production.

Source: prepared by Markestrat based on ANP and CitrusBR.

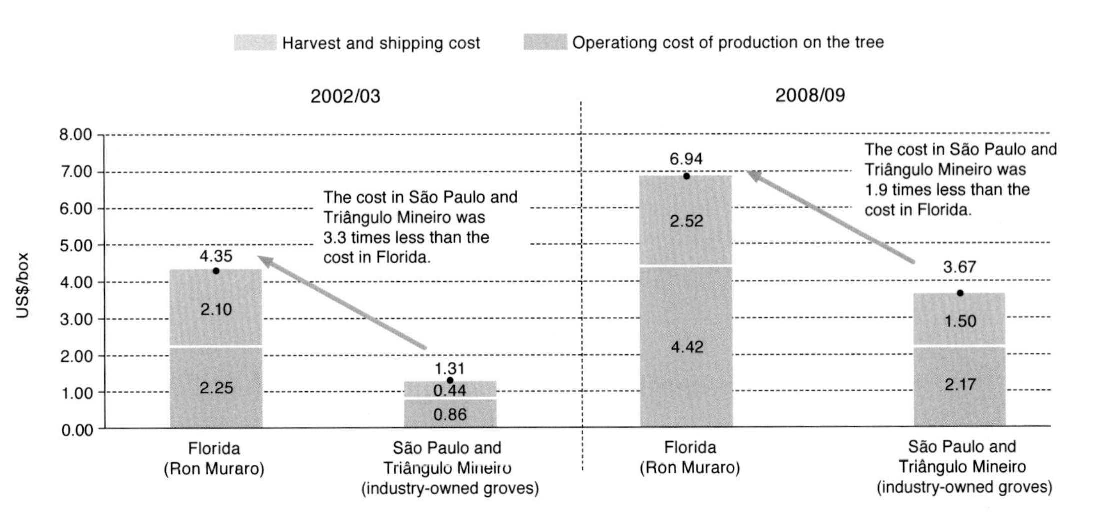

Graph 21. **Comparison of operating costs of orange production among industry-owned groves in Brazil's citrus belt and operating costs of orange production in Florida, in US$/box, between 2002/03 and 2008/09.**

Source: CitrusBR and production cost per acre/hectare: UF/IFAS – Citrus REC. Production cost per box calculated by dividing the production cost per acre/hectare of UF/IFAS, Ronald Muraro (06 March 2010) by the average agricultural productivity in Florida. Harvest and freight costs in 2002/03 from the Florida Citrus Outlook 2002/03 Season.

Table 18. History of production cost of oranges destined for the juice industry in central Florida.

	1982/83	1987/88	1992/93	1997/98	2002/03	2008/09
Productive acres in Florida (× 1000)						
Orange	536.8	380.2	489.2	609.2	587.6	459.1
Temples	15.8	9.3	7.3	6.2	4.2	0
Total	552.6	389.5	496.5	615.4	591.8	459.1
Productive hectares in Florida (× 1000)						
Orange	217.2	153.9	198	246.5	237.8	185.8
Temples	6.4	3.8	3	2.5	1.7	0
Total	223.6	157.6	200.9	249	239.5	185.8
Commercial productive trees in Florida						
Orange + temple	42.928	35.537	55.642	78.587	76.494	60.753
Density of productive trees (trees per acre and hectare)						
Acre	77.7	91.2	112.1	127.7	129.3	132.3
Hectare	192	225.5	276.9	315.6	319.4	327
Production in Florida (millions of boxes)						
Orange	139.6	138	186.6	244	203	162.4
Temples	4.7	3.6	2.5	2.3	1.3	0
Total	144.3	141.6	189.1	246.3	204.3	162.4
Agricultural productivity in Florida (boxes per acre, per hectare, and per tree) orange + temple						
Boxes/acre	261.1	363.5	380.9	400.2	345.2	353.7
Boxes/hectare	645.3	898.3	941.1	989	853.1	874.1
Boxes/tree	3.36	3.98	3.4	3.13	2.67	2.67
Total cost of production on the tree						
US$ per acre	548	628	779	766	778	1,566
US$ per hectare	1,354	1,551	1,923	1,890	1,922	3,866
US$ per box	2.10	1.73	2.04	1.91	2.25	4.42

Sources: Productive acres and hectares, productive commercial trees and total production in Florida: FDOC citrus reference book.

Agricultural yield in Florida: calculated by dividing the data of total production by productive acres, hectares and feet, both from the FDOC citrus reference book.

Production cost per acre/hectare: UF/IFAS – Citrus REC, presentation by Ronald P. Muraro on 06 March 2010.

Production cost per box: calculated by dividing the production cost per acre/hectare of the UF/IFAS – Ronald P. Muraro by the average Agricultural Productivity in Florida.

Harvesting and shipping costs in 2002/03: Florida Citrus Outlook 2002-03 Season.

20. Pesticides in citrus farming

In 2009 there was a 7.7% increase in sales of pesticides compared to the previous year, totaling 725,577 tons of commercial product, equivalent to the marketing of 335,816 tons of active ingredient. Of this total, 4.2% of sales of commercial products, equivalent to 5.7% of the active ingredients, were consumed by the citrus sector, accounting for a total of R$ 201 million (Graph 22). From 2008 to 2009, the sector reduced the consumption of pesticides by just over 20% (Graph 23).

Of the classes of pesticides, acaricides are most prominently used in citrus farming (representing 1.7% of sales in the pesticide sector in 2009), accounting for 88% of the overall value traded in 2009. Out of the total quantity of active ingredient consumed by the citrus industry, acaricides represent 39%, followed by foliar insecticides at 29%, and foliar-application fungicides at 14%. These three classes accounted for 55% of spending on pesticides in the citrus sector. The growing pressure of citrus greening and CVC has increased the consumption of citrus pesticides exponentially and from 2003 to the present day there has been an increase of around 600%.

In 2009, the citrus sector was the second most intensive crop in the use of pesticides. In all, 17.5 kg/ha of active ingredient were applied, of which 6.8 kg/hectare were acaricides and

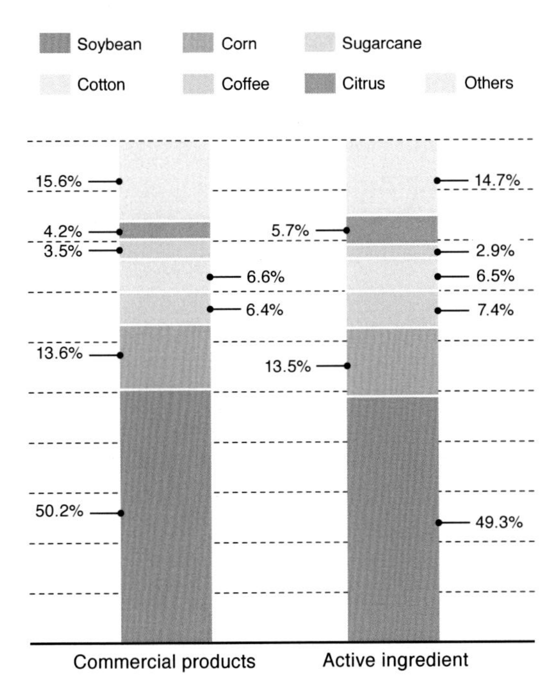

Graph 22. Share of crops in consumption of pesticides in Brazil – 2009.

Source: prepared by Markestrat based on SINDAG.

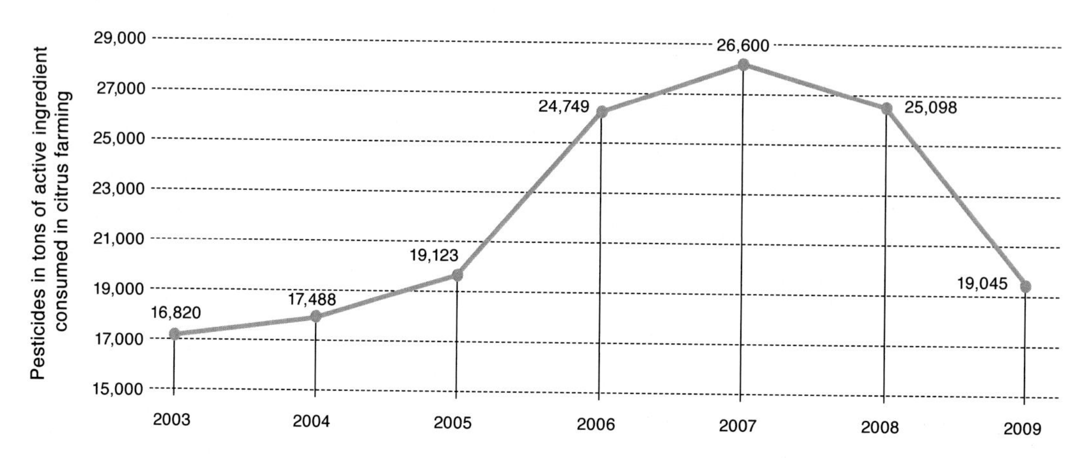

Graph 23. **Evolution in pesticide consumption in citrus farming.**
Source: prepared by Markestrat based on SINDAG.

5.1 kg/hectare were insecticides. Cotton came in first place (27.1 kg/hectare) and soybeans third (7.6 kg/hectare).

The expectation for the sector is for sales of pesticides in 2010 to exceed 2009 sales by 10%. For the citrus industry, the outlook for the 2010/11 season is an increase in consumption, due to the improved exchange ratio and more attractive prices for oranges and orange juice on the international market.

21. Use of fertilizers in citrus growing

Although there has been an overall decline of 1% in the quantity of fertilizer used throughout the agribusiness in relation to the previous year, in orange farming there was a drop of around 6.3%, reflecting the difficulty faced by growers in recent years, mainly due to the unattractive price and difficult access to credit. Nevertheless, the share in consumption remained stable at 2% of total consumption, behind 11 other crops (Graph 24).

In terms of consumption per unit (kg/ha), orange is ranked sixth in terms of use, with an application of 362 kg/ha in 2009, a decrease of 10.2% compared to 2008 and 26.3% compared to 2007. The share of the cost of fertilizer in gross revenue from the sale of oranges has also risen from 5% in 2007 to 7% in 2008 and 8% in 2009. This highlights the worsening in the exchange ratio for the period in this activity. In 2007, it took 60 boxes (40.8 kg each) of oranges to acquire one ton of fertilizer. In 2009, this figure rose to 95 boxes (Graph 25).

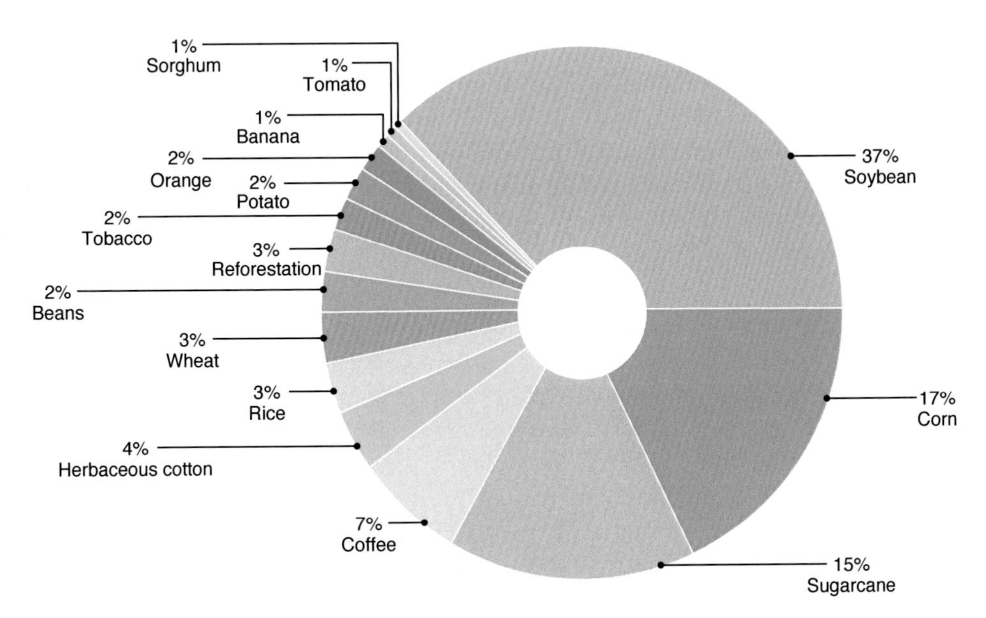

Graph 24. Share (percentage) in fertilizer consumption by crop.

Source: prepared by Markestrat based on ANDA data.

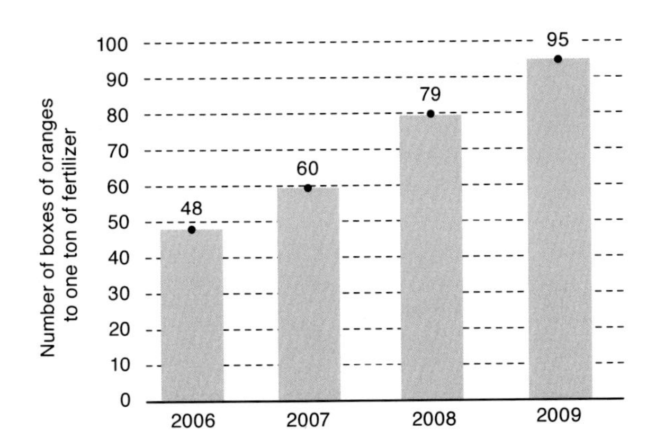

Graph 25. Exchange ratio between fertilizer and boxes of oranges.

Source: prepared by Markestrat based on ANDA data.

22. Minimum wage

This constitutes another major impact on the citrus sector because it is labor-intensive, especially in the harvesting stage, which is still done almost entirely by hand.

Therefore, the increases in the minimum wage tend to be a burden on the costs of production, reducing the profit margins in this activity.

In October 1994 – shortly after the start of the "Plano Real" that introduced Brazil's new currency – the minimum wage was R$ 70 and a box of oranges destined to the juice industry was quoted at R$ 2.92. The last minimum wage increase, in January 2010, brought it up to R$ 510, an increase of 628% in the period, while the average price of a box of oranges destined to the juice industry was quoted at R$ 7.70, representing an increase of 253% (Graph 26).

In the last five years, 2009 was the year with the worst exchange ratio between the price of a box of oranges and the minimum wage, i.e. in that year, the amount of the monthly minimum wage was the equivalent of 92 boxes of oranges, 49 boxes more than in 2008. The recovery in the price of a box of oranges in the first half of 2010 helped reduce the impact on production costs, but the exchange ratio of 48 boxes to the amount of the monthly minimum wage remains well above that of October 1994.

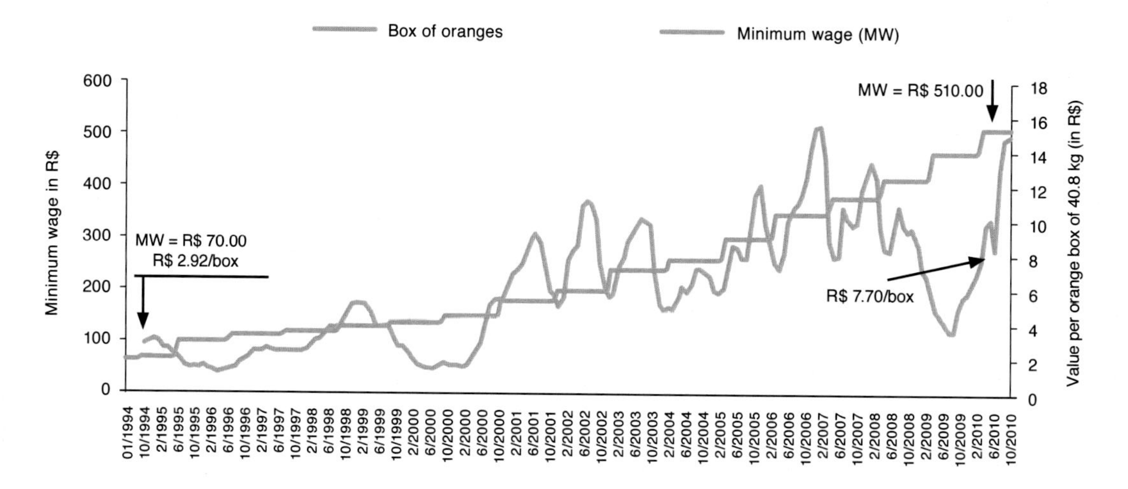

Graph 26. Evolution of the price of a box of oranges on the spot market (CEPEA) versus minimum wage.

Source: prepared by Markestrat based on CEPEA and Ipeadata.

23. Jobs and working conditions

In the São Paulo citrus belt, as in the major orange-producing regions around the world (Florida, California, Portugal, Spain, Italy, Greece, China and India), the activities of planting the orchards, cultivating the crops, harvesting the oranges and shipping the produce to the point of purchase, whether a packing house or a juice factory, are the responsibility of growers, not unlike other fruits and commodities cultivated worldwide, such as apples, grapes, coffee, soybeans, corn, wheat, and others.

According to the Ministry of Labor and Employment, the 2009/10 growing season started in July with roughly 58,000 workers in Brazil engaged in orange farming. Throughout the season, temporary workers are hired who are normally laid off at the end of the period. In July 2010, the number of workers hired since the beginning of the harvest totaled 94,000. Thus, during the 2009/10 growing year, around 150,000 workers were allocated to field activities, while in the concentrated orange juice industry there are roughly 7,000 workers permanent workers and 4,000, temporary workers, totaling 11,000 during the growing year. In June 2010, the overall balance of workers in orange farming was 77,000, and around 7,000 in the orange juice industry. Considering that, in the orange production chain, each direct job in the field generates two indirect jobs along the chain, it is estimated that there are roughly 230,000 workers involved in the citrus sector.

In terms of remuneration, according to data from the Labor Ministry's Annual Report of Social Information, workers allocated to orange juice factories, who generally have higher education levels than workers in the field, are better paid, i.e. on average they earn about R$ 1,445.00 a month, versus R$ 680 a month for workers engaged in cultivation. The total amount of wages generated in the citrus sector in the 2009/10 agricultural year is estimated at R$ 676.62 million, or US$ 378.4 million (Graph 27).

No other crop absorbs such a high number of temporary workers per hectare in the state of São Paulo as citrus. On sugarcane croplands, for example, the ratio is one temporary job for every 41 hectares, while in the case of oranges, the ratio is one temporary job for every 9 hectares. This figure shows the importance of the citrus sector in generating jobs in the field, helping to drive the economy of many municipalities, mostly located in the state of São Paulo. In the harvesting activity, there is no distinction between men and women in hiring, the restriction is that workers must be over 18 years old. The teams of orange pickers in 2009/10 were comprised of 65% men and 35% women. In sugarcane, 90% of the workers are men.

Occupational health and safety standards are guaranteed to workers by Brazilian law, namely Regulatory Standard 31, which establishes the requirement for agricultural employers to provide appropriate working conditions, hygiene and comfort. Accordingly, a number of items must be covered, such as the use of suitable clothing, hats or caps to protect against sunburn, footwear to protect workers against insects and weeds, availability of restrooms and shelter from the rain, as well as suspending the harvest during rainstorms.

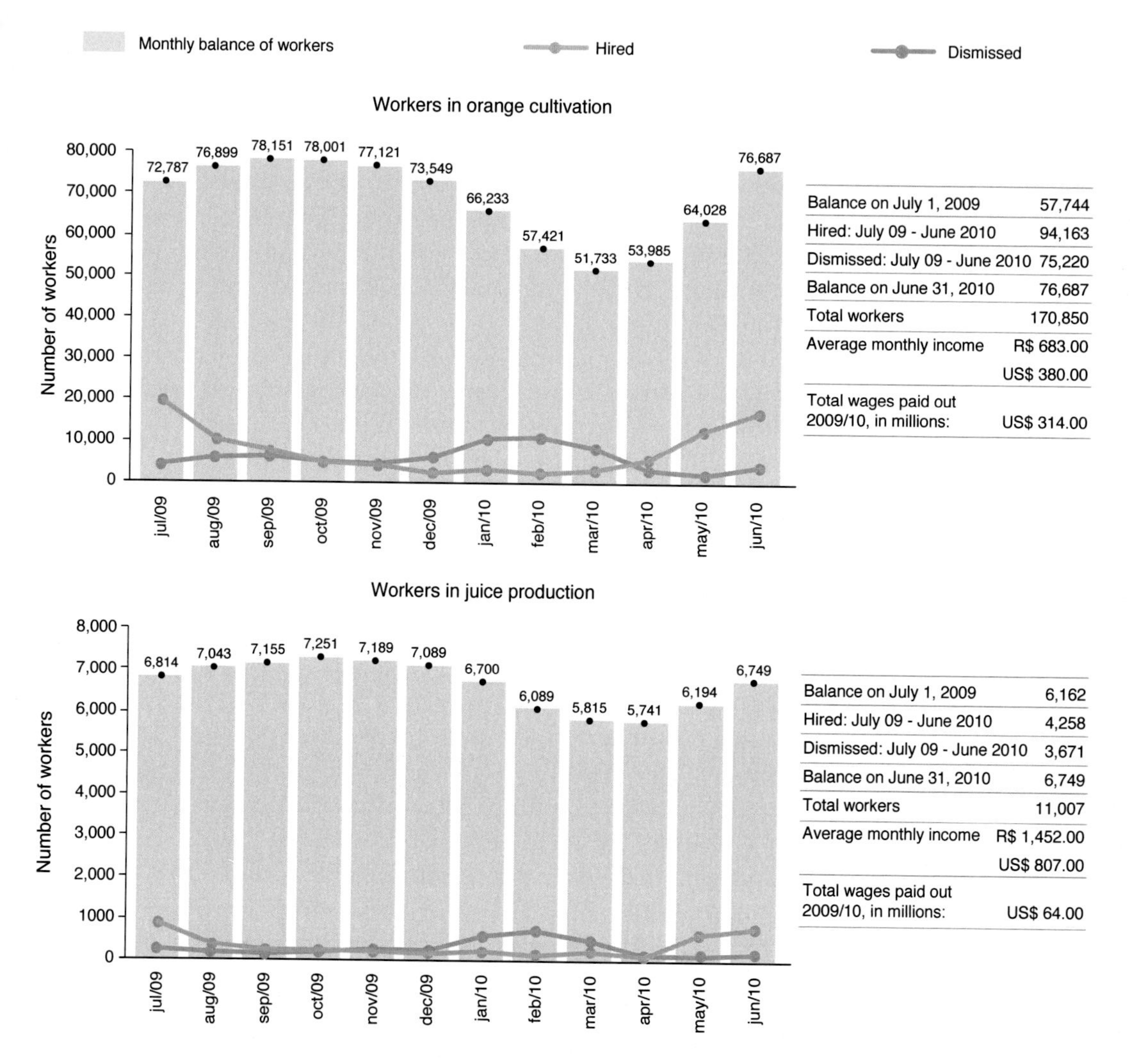

Balance on July 1, 2009	57,744
Hired: July 09 - June 2010	94,163
Dismissed: July 09 - June 2010	75,220
Balance on June 31, 2010	76,687
Total workers	170,850
Average monthly income	R$ 683.00
	US$ 380.00
Total wages paid out 2009/10, in millions:	US$ 314.00

Balance on July 1, 2009	6,162
Hired: July 09 - June 2010	4,258
Dismissed: July 09 - June 2010	3,671
Balance on June 31, 2010	6,749
Total workers	11,007
Average monthly income	R$ 1,452.00
	US$ 807.00
Total wages paid out 2009/10, in millions:	US$ 64.00

Graph 27. **Mapping of workers in orange cultivation and juice production.**

Source: prepared by Markestrat based on data from RAIS and CAGED.

24. Production incentive policies

Despite the economic and social importance to Brazil, the citrus sector needs policies to boost production, mainly in the segment of small-scale farmers, who have suffered because they have failed to renew their orange groves with more appropriate technology for improving productivity.

In the Agriculture Ministry's "Agricultural and Livestock Plan" for the 2010/2011 season, for example, no specific guidelines are included for the sector regarding topics such as: Minimum Price, Special Line of Credit (SLC) and Rural Credit for Funding and Marketing. For citrus farming, there is also no direct mention of the limits of funding advances and loans from the Federal Government, falling under the guidelines of the more comprehensive sector of "Fruit-Growing."

In the area of risk management, citrus fruits are one of the crop types covered by the Climate Risk Agricultural Zoning Program Studies and are included in the Rural Insurance Premium Subsidies Program. However, the latter is still only rarely used, reaching only 11% of the total area occupied by Brazilian agriculture in 2009.

Some improvements in terms of citrus production incentives were observed, albeit timidly, in the 2010/11 growing season. Citrus-farming enterprises were gradually covered by the Program for Guaranteed Agricultural Activity (Proagro), which aims to release growers from their financial obligations in rural funding credit operations and indemnify the growers own resources applied in the case of financial losses resulting from adverse weather events or pests and diseases with no widespread method of combat, control or prevention.

Through an initiative by the São Paulo state secretary of agriculture, João de Almeida Sampaio Filho, a major breakthrough in the sector was the creation of rural insurance against citrus canker and greening, contributing up to R$ 35 million in resources. Citrus growers with up to 20,000 trees will benefit from the program, i.e. around 87% of the growers in the state of São Paulo. New incentives also include the donation of 36 juice-extracting machines for making freshly squeezed orange juice, to the municipal governments in the citrus-growing regions interested in including orange juice on the lunch menu at public schools.

Although these recent achievements are relevant to the sector, there are still major barriers to overcome, such as the issue of taxes. Orange juice for sale on the domestic market is heavily taxed when compared to sales on the foreign market, discouraging the expansion of consumption on the domestic market; taxes levied on revenues from sales alone are as follows: ICMS (VAT) – 12%; IPI – 5%; PIS – 1.65%; and COFINS – 7.60%. Moreover, citrus-exporting companies cannot use presumed PIS and COFINS credit, i.e. credit on purchase of raw materials acquired from individuals. These federal taxes initially established in Article 8 of Laws 10637 and 10833 were eligible for offset against the PIS and COFINS taxes themselves with other taxes, or even subject to reimbursement. Then, Law 10925 was passed, which repealed Article 8 of both of the aforementioned laws, and began to deal with the presumed credit, still subject to offset against the PIS and COFINS taxes. However, this law no longer stipulated the possibility of offset with other taxes or being subject to reimbursement. In December 2005, the Receita Federal (Brazil's IRS) published Interpretative Declaratory Act No. 15 in the Official Journal of Brazil (Diário Oficial), affirming that such credits could not be offset with other taxes or reimbursement. In March 2006, the Receita Federal issued the Normative Instruction No. 636, to regulate the presumed credit, confirming the impossibility of offsetting against other taxes or reimbursement, effective as of August 2004, i.e. the effects of the publication of Law 10925.

Hence, predominantly exporting companies cannot use this tax credit, since it can only be offset against the PIS and COFINS taxes themselves, accumulating a significant non-usable amount. It is estimated that this amount is somewhere around R$ 0.30 per box, which could return somewhere in the region of R$ 60 million to the growers, assuming sales of 200 million boxes a year. The estimated cumulative stock of federal tax credits (PIS and COFINS) for the sector reached roughly R$ 450 million in October 2010. In relation to the stock of ICMS, the estimate for the sector is around R$ 260 million up to the same month.

Moreover, in developed countries and some emerging countries, there is well-known support for the agricultural sectors, whether in the form of mechanisms of aid to growers or direct subsidies. According to a recent study by Professor André Nassar of the ICONE Institute, considering all the types of subsidies granted to farmers, there is significant differentiation of Brazil in relation to the USA, the European Union, and Japan. Whereas American farms receive an average of US$ 56,000 per year, European farms receive around US$ 27,000; Japanese farms, around US$ 20,000; while Brazilian farms receive US$ 1,100. Calculating the total amount of subsidies in relation to the total value of production, the result is 63% in Japan, 43% in the USA, 33% in the European Union and just 6% in Brazil.

The deliberate policy of a number of developed nations and some emerging nations, such as India, to transfer revenues from the urban economy to the rural economy is not implemented in Brazil. The Brazilian consumer benefits from agricultural products at market prices. The tax payer does not have to bear the cost of revenue problems in the agricultural sector, as seen in the case of the developed nations. The Brazilian model is better, although its products have to face unfair competition from the subsidized farming of other countries. Even though industrial manufacturing sectors such as the automobile and white goods industries have received temporary incentives from the Brazilian government with the reduction of the IPI (Industrialized Goods Tax), the agricultural sector has at no time enjoyed such benefits.

The cases of subsidies for cotton in the USA and for sugar in the European Union, along with the use of anti-dumping measures against Brazilian orange juice in the USA, are recent examples of the difficulties faced by Brazilian exporters.

This is an infinite agenda, in which the Brazilian government, together with the private sector and supported by universities and research institutes, needs to act with rigor and efficiency.

25. Cycle of working capital and available funding sources

The orange juice companies are among those most heavily penalized by the long cycles and slippage between the financing of harvests – due to the disbursement of funds to the suppliers of raw materials – and the receivables from international customers. The need for working capital in the juice industries varies from 9 to 11 months and the larger their stocks between one harvest and the next, the greater their need for working capital.

In contrast to cattle, which is supplied to the slaughterhouse throughout the year, processed and exported, because it is seasonal, oranges only reaches the industries in the second half of the year. This being the case, all the oranges delivered to the packers from the seaport terminals in Belgium, the Netherlands, United States, Japan and South Korea in the period from January to September are oranges paid for by the industry and processed between July and December of the preceding year. Therefore, due to the advance payments made to the citrus farmers before the start of the harvest, and the grace periods for payment granted to the packers following delivery of the physical product overseas, the Brazilian industry ends up taking on an important role in financing the production throughout the productive chain.

The strengthening of the government's credit activities would be significant, especially for those producers that direct their production at fresh consumption, or for greater freedom for the producers to negotiate their production with the industry.

According to the Brazilian Statistical Yearbook for Rural Credit, published by the Brazilian Central bank, in 2009 the national financial system granted R$ 75 billion to fund farming through 2.5 million contracts. Of this amount, R$ 54 billion was used for agriculture, with 47% being used to cover costs, 19% for commercialization (EGF Federal Government Loans, pre-commercialization, CPR (Rural Credit Note) and discounts on NPR (Rural Promissory Note) and DR (Rural Trade Bill)), 2% for investment, 1% for processing and industrialization and 1% for stockpiling, among other items. However, only a small percentage of the funding is applied to citriculture.

Of the funds allocated to cover costs, 3% went to citrus farming, versus 32% for soy, 17% for corn, 11% for coffee and 9% for sugarcane; of the funds allocated to commercialization, 0.1% was destined for citrus farming, versus 25% for corn, 17% for rice, 10% for sugarcane, 6% for soy and 3% for coffee. Of the funding allocated to processing or industrialization, citriculture received less than 0.1% of the volume, versus 44% for sugarcane and 41% for coffee; and for the purposes of investment to establish perennial farming 8% was applied to citrus groves, versus 33% for sugarcane and 10% for coffee. The granting of credit for stockpiling is made difficult by the high perishability of citrus fruit, which prevents it from being stocked for extended periods.

Citriculture was allocated a total of R$ 948.5 million by way of 13,853 contracts with an average value of approximately R$ 68,500; with 95% of the volume being applied to orange farming. The financial resources were distributed via the concession of cost covering (R$ 855 million); investment for the establishment of citrus groves (R$ 85.5 million); funding for commercialization, via discounts in NPR (Rural Promissory Note) and DR (Rural Trade Bill)

(R$ 5.4 million); CPRs (Rural Credit Notes) (R$ 2.5 million); and funding for processing and/or industrialization (a single contract at a value of R$ 150,000). It should be pointed out that the CPRs (Rural Credit Notes) mentioned here refer solely to those registered on the national financial system, and do not include the over-the-counter CPRs (Rural Credit Notes), which play an important role as an instrument of finance.

São Paulo was the State that received most funding released for citriculture, with around 90% of the funds available for covering production costs and 54% of the funding for investment. Also, 100% of the financial volume of the CPRs (Rural Credit Notes) was destined for orange groves in the State of São Paulo. The State of Bahia was allocated 16% of the funds destined for investment in citrus groves. However, while the average contract value in São Paulo was R$ 133,000, in Bahia the figure was R$ 22,000. Citriculture is definitely worthy of more attention.

26. Price of oranges

As in the case of the other agricultural commodities, both the citrus farmer and the processing industry are subject to price fluctuations, depending on the variation between the demand for oranges, the expectations for consumption on the world juice market, domestic consumption of fresh fruit and the forecasts for stocks between harvests. In general, the price received by the producer varies according to the destination of the fruit, given that the value aggregated to oranges for fresh consumption tends to be higher than for oranges sent for industrial processing, due to the stricter requirements in terms of visual and intrinsic aspects (Graph 28).

Brazilian orange juice industries plant an average of 35% of the fruit they use for the production of juice, acquiring 65% of their raw material from independent producers, who are free to plan the destination of their production, choosing to sell to industry or to the market for fresh fruit.

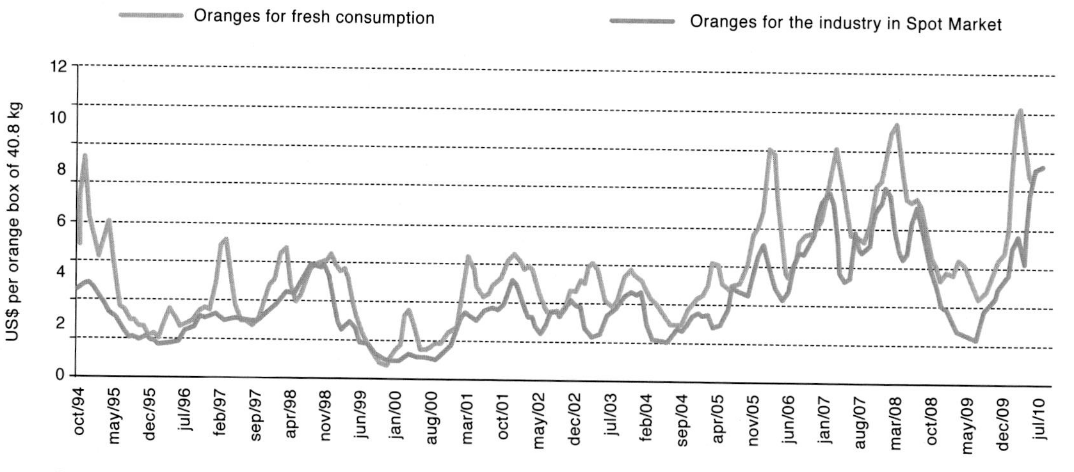

Graph 28. Evolution of the price paid to the producer per box of oranges.
Source: prepared by Markestrat based on the Cepea.

The citrus farming chain in the State of São Paulo works similarly to the one in Florida. The industry buys the oranges from the citrus farmers using a number of models, such as, long-term contracts with pre-determined fixed prices; long-term contracts with or without a minimum guaranteed price and with price triggers indexed to the real audited averages, obtained from the selling prices for the concentrated juice to each of industries on the overseas market between the period of June and July for each harvest; long-term contracts with or without a minimum guaranteed price directly linked to the daily and average annual prices for the commodity on the New York Stock Exchange; contracts to buy oranges during the harvest period at the price for the day, the so-called spot market or gateway; and also via long-term leasing or agricultural partnership contracts.

The prices for oranges in each mode are determined by the state of supply and demand for orange juice at the time when each contract is signed. The supply of and demand for juice and oranges on the market is based on the prices for juice quoted on the New York Stock Exchange. These prices and the other market conditions may vary within a single harvest, generate prices for contracts to buy oranges that differ from each other depending on the market conditions at the time when each contract is signed.

Each type of contract has its own inherent advantages and risks. Long-term contracts, generally with a lifespan of two to five years, protect the producer against negative fluctuations in the price for juice, as was the case in the harvests for 2007/08, 2008/09 and 2009/10. However, in these cases, the producer does not take advantage of the market opportunities when there is a rise in the price on the spot market, as witnessed in the first half of 2010. The spot market, in turn, is unpredictable, reflecting the specific market conditions for each harvest. When orange juice is at a high, those selling on the spot market get better prices than the ones for long-term contracts. When the international market is at a low, it is generally the producers with fixed-price contracts that obtain the best result.

The price paid by industry for oranges results in current and future international prices for juice, as well as market expectations regarding the future supply of and demand for oranges, at the time in which each contract for buying oranges is negotiated. Graph 29 shows that there is a relative cohesion between the price of orange juice on the New York Stock Exchange and the price of a box of oranges on the spot market in Brazil.

The price of a box of oranges directly affects the cost of production for Brazilian orange juice, being a determining factor in its degree of competitiveness in relation to other beverages, such as the juices of other fruits and other non-alcoholic beverages in general. Another factor to be considered in the competitiveness is the import duty paid in the United States for the entry of Brazilian orange juice and the distribution and port costs placed on the Brazilian product to be transported to that destination. The prices applied in the harvests from 2002/03 to 2008/09 reveal that the values received by the Brazilian citrus farmers were close to those received by producers in Florida (Table 19).

26. Price of oranges

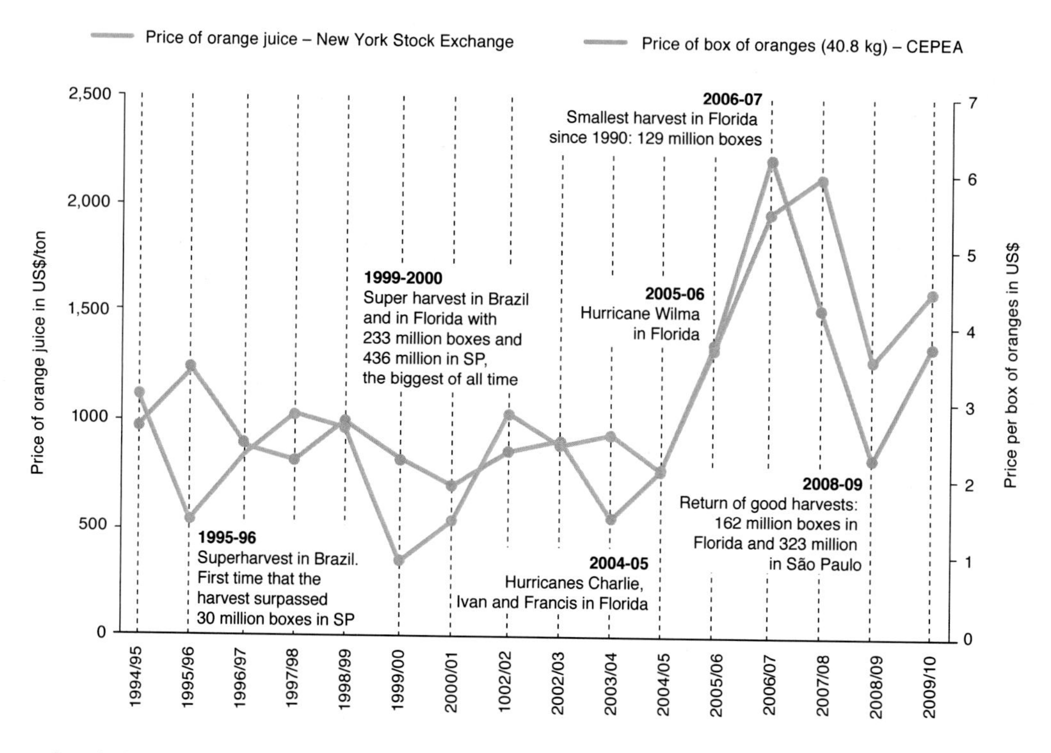

Graph 29. Price of FCOJ on the New York Stock Exchange versus price of box of oranges.
Source: prepared by Markestrat based on CitrusBR and CEPEA.

Table 19. Comparison between the price per box of orange in Brazil's citrus belt and in Florida.

	2002/03	2003/04	2004/05	2005/06	2006/07	2007/08	2008/09	Average
Revenues for orange on tree (supplied to the industry in Florida)								
Early-Sprouting Varieties	US$ 2.42	US$ 2.09	US$ 2.54	US$ 4.60	US$ 8.70	US$ 5.79	US$ 4.60	US$ 4.39
Valencia	US$ 3.80	US$ 3.67	US$ 4.17	US$ 6.38	US$ 11.55	US$ 7.30	US$ 6.25	US$ 6.16
Total	US$ 3.05	US$ 2.85	US$ 9.91	US$ 5.49	US$ 10.12	US$ 6.57	US$ 5.40	US$ 5.26
Exchange rate								
July	R$ 2.90	R$ 2.87	R$ 3.04	R$ 2.37	R$ 2.19	R$ 1.89	R$ 1.59	
August	R$ 3.13	R$ 3.00	R$ 3.01	R$ 2.36	R$ 2.16	R$ 1.96	R$ 1.61	
September	R$ 3.31	R$ 2.92	R$ 2.90	R$ 2.30	R$ 2.16	R$ 1.98	R$ 1.79	
October	R$ 3.82	R$ 2.86	R$ 2.85	R$ 2.26	R$ 2.15	R$ 1.81	R$ 2.17	
November	R$ 3.59	R$ 2.91	R$ 2.79	R$ 2.21	R$ 2.15	R$ 1.76	R$ 2.25	
December	R$ 3.63	R$ 2.93	R$ 2.72	R$ 2.28	R$ 2.15	R$ 1.79	R$ 2.39	
January	R$ 3.43	R$ 2.85	R$ 2.69	R$ 2.28	R$ 2.14	R$ 1.77	R$ 2.31	

Table 19. **Continued.**

	2002/03	2003/04	2004/05	2005/06	2006/07	2007/08	2008/09	Average
CEPEA indication for the price of oranges delivered to factory for the spot mode (reals per 40.8 kg box)								
July	R$ 7.75	R$ 7.85	R$ 5.51	R$ 8.71	R$ 10.06	R$ 10.93	R$ 10.95	
August	R$ 8.25	R$ 8.75	R$ 6.22	R$ 8.44	R$ 10.76	R$ 10.16	R$ 9.71	
September	R$ 8.48	R$ 9.24	R$ 5.98	R$ 7.94	R$ 11.04	R$ 9.78	R$ 9.33	
October	R$ 10.85	R$ 9.72	R$ 6.39	R$ 7.86	R$ 11.52	R$ 9.89	R$ 9.57	
November	R$ 11.21	R$ 10.20	R$ 7.23	R$ 9.70	R$ 12.51	R$ 11.77	R$ 8.63	
December	R$ 10.98	R$ 9.98	R$ 7.31	R$ 11.53	R$ 14.26	R$ 12.61	R$ 7.27	
January	R$ 10.07	R$ 9.87	R$ 7.08	R$ 12.13	R$ 15.46	R$ 13.46	R$ 6.80	
Average CEPEA spot price delivered to factory	R$ 9.66	R$ 9.37	R$ 6.53	R$ 9.47	2$ 12.23	R$ 11.23	R$ 8.89	R$ 9.63
Minus cost of picking	-R$ 0.84	-R$ 1.09	-R$ 1.06	-R$ 1.27	-R$ 1.41	-R$ 1.52	-R$ 1.91	-R$ 1.29
Minus cost of transportation to factory	-R$ 0.66	-R$ 0.74	-R$ 0.81	-R$ 0.88	-R$ 0.90	-R$ 0.89	-R$ 1.07	-R$ 0.85
Average spot price estimated on the tree	R$ 8.16	R$ 7.60	R$ 4.66	R$ 7.32	R$ 9.92	R$ 8.82	R$ 5.91	R$ 7.49
CEPEA indication for the price of oranges delivered to factory for the spot mode (dollars per 40.8 kg box)								
July	US$ 2.67	US$ 2.73	US$ 1.81	US$ 3.68	US$ 4.59	US$ 5.79	US$ 6.87	
August	US$ 2.64	US$ 2.92	US$ 2.07	US$ 3.58	US$ 4.99	US$ 5.17	US$ 6.05	
September	US$ 2.56	US$ 3.16	US$ 2.06	US$ 3.44	US$ 5.10	US$ 4.95	US$ 5.20	
October	US$ 2.84	US$ 3.39	US$ 2.24	US$ 3.48	US$ 5.36	US$ 5.48	US$ 4.42	
November	US$ 3.13	US$ 3.51	US$ 2.59	US$ 4.39	US$ 5.81	US$ 6.68	US$ 3.83	
December	US$ 3.03	US$ 3.41	US$ 2.68	US$ 5.07	US$ 6.63	US$ 7.06	US$ 3.04	
January	US$ 2.93	US$ 3.46	US$ 2.63	US$ 5.32	US$ 7.23	US$ 7.59	US$ 2.94	
Average Cepea spot price delivered to factory	US$ 2.83	US$ 3.23	US$ 2.30	US$ 4.14	US$ 5.67	US$ 6.10	US$ 4.62	US$ 4.13
Minus cost of picking	-US$ 0.25	-US$ 0.35	-US$ 0.37	-US$ 0.55	-US$ 0.65	-US$ 0.83	-US$ 0.96	-US$ 0.57
Minus cost of transportation to factory	-US$ 0.19	-US$ 0.2	-US$ 0.28	-US$ 0.38	-US$ 0.42	-US$ 0.48	-US$ 0.54	-US$ 0.36
Average spot price estimated on the tree	US$ 2.39	US$ 2.63	US$ 1.65	US$ 3.21	US$ 4.60	US$ 4.79	US$ 3.12	US$ 3.20
Additional cost for american import duty	US$ 1.79	US$ 1.79	US$ 1.79	US$ 1.79	US$ 1.79	US$ 1.79	US$ 1.79	US$ 1.79
Average CEPEA spot price estimated on the tree with American import duty	US$ 4.18	US$ 4.42	US$ 3.44	US$ 5.00	US$ 6.39	US$ 6.58	US$ 4.91	US$ 4.99

Source: drawn up by Markestrat based on CEPEA and Citrus Reference Book 2010.

In this period, according to the Citrus Reference Book published in August 2010, the price paid to producers in Florida for the oranges on the tree was US$ 5.26 per box, while in Brazil the producer was paid an average of US$ 4.13 on the spot market, delivered to factory measured by the CEPEA.

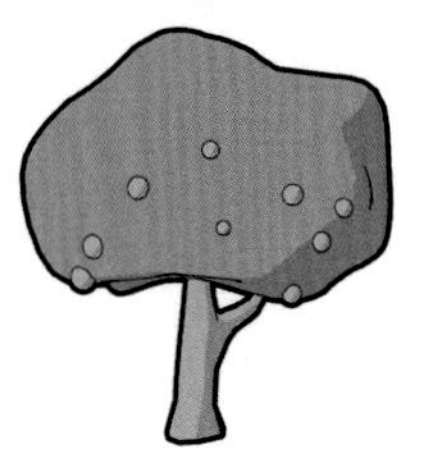

Florida US$ 5.26
Price on the tree

Brazil US$ 4.13
CEPEA Spot price
delivered to factory

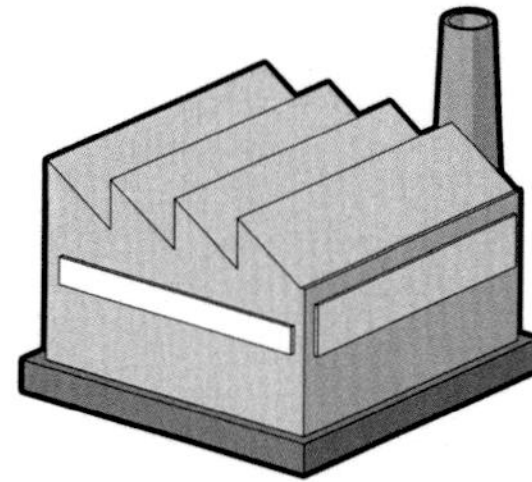

However, given the fact that the Citrus Reference Book only publishes the price on the tree and the CEPEA only reports the price delivered to factory, it is necessary to subtract the cost of picking and transportation to the industry from the CEPEA price, which in accordance with the compilations made by the international auditing firm, on average, in the period from 2002/03 to 2008/09, was US$ 0.57/box for picking and US$ 0.36/box for transportation, thereby resulting in an estimated CEPEA on the tree cost of US$ 3.20.

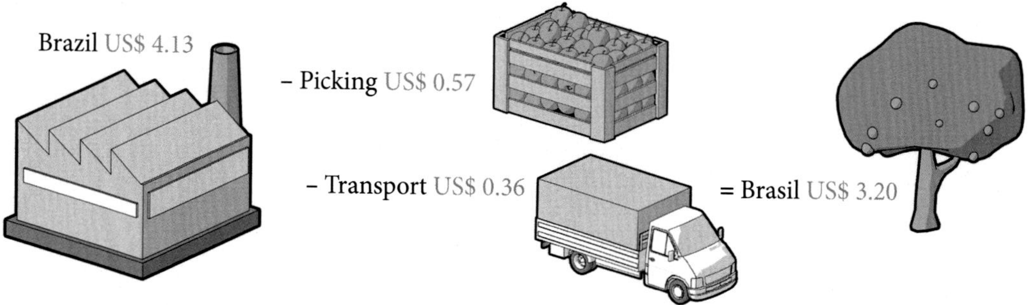

After this, it is also necessary to add US$ 1.79/ box for the American Import Duty on juice, which is paid for the product to enter that country, at an amount equivalent to the box of oranges, this would be the equivalent to an on-the-tree price for Brazilian oranges of US$ 4.99 per box (after payment of the American "toll"), just 5% lower than the price received by the producer in Florida. If consideration is given to the logistical costs for transporting the juice from the industry in the São Paulo state to Florida, estimated at US$ 180 per ton of FCOJ or US$ 0.77/box of oranges, one reaches the conclusion that the box of oranges would have cost the Brazilian industry US$ 5.76, equivalent to 9.5% higher than the amount received by the producer in Florida and, therefore, more than what oranges cost to an industry in Florida. This analysis must also take into consideration the fact that the average yield of juice in fruit from the citrus belt was 6.6% lower than that from Florida in the same period, which undermines the Brazilian industry even more.

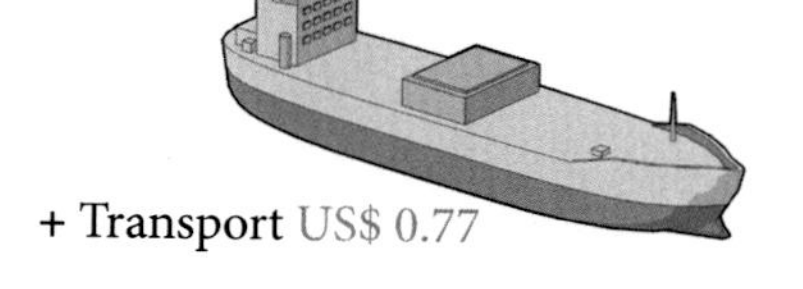

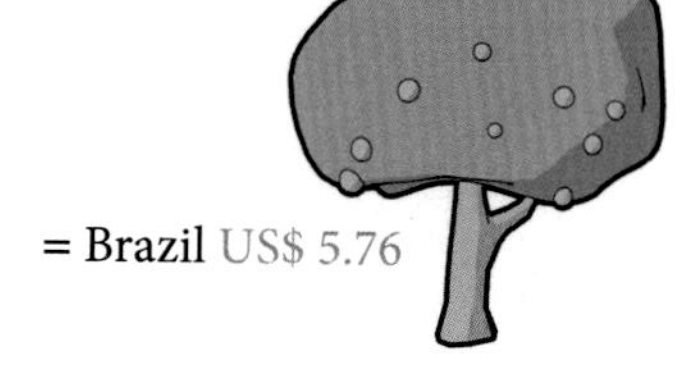

The consecitrus

Orange producers and juice industries are working together to establish a viable price-referencing mechanism called CONSECITRUS, inspired by the model for the CONSECANA, created in conjunction with the sugarcane producers and the sugar mills, which has been working successfully for a number of years. The Department of Agriculture for the State of São Paulo, together with entities representing the citrus farmers and CitrusBR, which represents the orange processing industries, have focused efforts to this end and are close to reaching a definition for this new system.

The idea is to form a Board to represent the industries and the orange producers, which will establish the mechanism for identifying a reference price for a box of oranges. As in the case of the CONSECANA, the mechanism for identifying the reference price for oranges in the CONSECITRUS could start with the selling price for Brazilian juice overseas, minus the average real costs of industrial production, logistics, international distribution and sale of the juice.

These costs would be identified by an independent international auditing firm, respected by the parties involved.

To make the CONSECITRUS viable, the industry would provide information on the costs referred to above, on an individual basis and with a commitment to confidentiality, to the internationally certified independent auditing firm so that it could calculate the averages for these costs. It is this average that would be made available to the CONSECITRUS for the purposes of calculating the reference price for oranges, in conformity with the mechanism for identifying costs, for which the details are being negotiated. Since the work of the independent auditing firm was recently concluded, as reported by CitrusBR, the results has been released by the industry for publication, for the first time, in this survey.

Publication of these costs represents a milestone for the orange productive chain,, since the sector will now be able to see how much it costs for the Brazilian juice industry to process the oranges, to store, transport and sell the juice overseas, providing transparency to the overall procedure. The citrus farmer, in turn, will be able to negotiate the price of his oranges with the industry, based on information regarding the respective costs, which will lead to subsidies in the negotiation.

Via CitrusBR, the industry also confirmed that all the amounts for costs, the average figures for which were calculated by an independent auditing company, will also be confirmed at an opportune moment by the independent auditors for each of the industries, all at an international standard, in such a way that these numbers will be vested with credibility for the purposes of use by CONSECITRUS. This is the idea anyway!

Table 20 presents the average costs to the industry in 2009 for processing the oranges and transporting them, per ton of FCOJ at 66° Brix from the factory gates to the overseas port terminals.

Table 20. Average cost to industry per ton of FCOJ at 66° Brix from the factory gates to the overseas port terminals audited by international consulting firms.

Maritime logistics, port operations, administration, commercialization and sales costs and the funding of working capital overseas	US$ 158.39
Costs for overland shipping, port operations and tariffs in Brazil	US$ 79.16
Costs for processing oranges and the production of FCOJ and by-products, along with costs for administration and the funding of working capital in Brazil, subtracted from the revenues for the FOB factory by-products	US$ 295.81
Total cost to Brazilian industry for processing and logistics from the factory gates to the port terminals in Europe, excluding amortization and depreciation expenses and the cost of capital invested in orange processing and transportation of FCOJ	US$ 533.36

The costs listed above obey a specific methodology for analyzing the average costs of operations in the industrial sector that exported citrus juices in 2009. They may be subject to alterations resulting from changes in the cost of any of the sub-items calculated. Analysis of the costs was made in compliance with all applicable legal requirements regarding the conducting of business deals in the sector.
Source: CitrusBR.

27. Price of orange juice: an incredible volatility

The drop in the price for juice coincides with superharvests of oranges in São Paulo and/or in Florida. In such a scenario, expectations are for lower-priced oranges and, consequently, lower production costs for the juice. In this way, any news of larger than normal orange harvests leads to a drop in the prices for juice on the New York Stock Exchange and for the European buyers, with direct implications for the price to be paid for oranges by the industry. Inversely, the increases in prices for juice occur at times when the harvests fail, caused by the consequent expectation of increased prices for the oranges and higher production costs for the juice. So, any news regarding a reduction in the orange harvest brings about an increase in juice prices on the New York Stock Exchange, allowing the industry to negotiate new contracts for the sale of juice at higher prices to European importers.

A structural phenomenon that has also been negatively affecting the price of juice, and consequently the price of oranges, is the decrease in juice consumption, as a result of substitution by other low-calorie drinks such as flavored waters, sports drinks and others.

Therefore, the expectation of orange harvests and future consumption determines the price of orange juice on the New York Stock Exchange, affecting the selling prices of the juice by industries. The prices of juice, in turn, affect the prices of oranges to be paid by the industries. The supply of oranges is affected, among other factors, by the number of trees, productivity per orange tree, and climate; all of these factors are of an unpredictable nature, which can lead to significant fluctuations in production, whereas variation in the demand is lower. It's a complicated equation.

One adult orange tree in full bloom releases between 20,000 and 30,000 flowers. Of this total, only 2.5% to 4% survive, and after 11 to 13 months, the fruits reach the ideal point of ripeness and are harvested and shipped to be marketed as fresh fruit, or to orange juice processing industries.

It is this subtle oscillation in the bloom payment ratio that leads to variation of up to 30% from one season to another in the productivity index and total volume in any citrus belt around the globe.

After the hurricanes that struck Florida in 2004 and 2005, as shown in Graph 30, the price of orange juice underwent successive increases because of the reduction in supply in the state, which became insufficient to meet the demand. Since the price of a box of oranges follows the same trend as the price of juice on the international market, the successive price hikes also increased the price of the fruit. In 2006/07, the price of orange juice concentrate on the New York Stock Exchange reached a monthly record price of US$ 2.01/lb in December 2006. It was also in the aforementioned growing season that the price of a box of oranges reached higher values, being sold on the Brazilian spot market at an average price of R$ 12.00, up 30% compared to the price in the previous season.

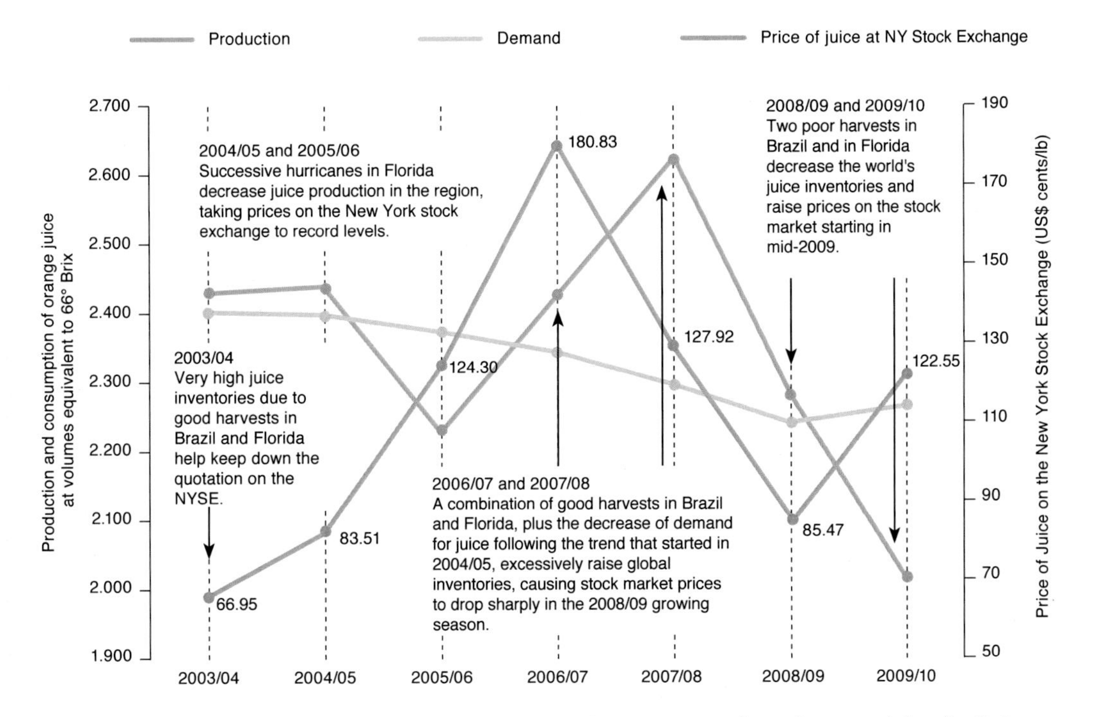

Graph 30. Comparative analysis between production, consumption of orange juice (both in quantities equivalent to 66° Brix) and the price of FCOJ on the NY Stock Exchange.

Source: prepared by Markestrat based on CitrusBR.

This appreciation occurred because Brazilian production had begun to sustain part of the US demand that was still suffering from declines in production. In the 2006/07 growing season, in the United States there was a 15% decrease, while in Brazil there was an 8% increase in the overall volume produced. It was a good time for the sector.

Once the major effects caused by the hurricanes in Florida had passed, with the relative recovery in US production, factors such as the global economic crisis and cheaper drinks available on the world market decreased American and European consumers' interest in orange juice. Brazilian exports of FCOJ and NFC converted to 66° Brix fell from the historical peak of 1.36 million tons in the 2006/07 growing season to 1.15 million tons in 2008/09. This 15.4% decrease caused the worldwide demand for Brazilian juice to fall by 209 tons, equivalent to a decrease in the annual demand of roughly 52 million boxes of oranges. The lower demand and growing supply toppled the price of orange juice.

This drop in the price of orange juice concentrate, in part, reached final consumers in Europe, but did not reach consumers in the United States, where retail prices remained stable and high, further accentuating the decrease in the consumption of orange juice. The fact that the price reduction for juice concentrate was not all passed on to the consumer increases the margin of the bottlers with high bargaining power, since practically 35 bottlers buy roughly 80% of the global volume. The bargaining power of bottlers along with higher supply at volumes greater than the demand in certain crops tends to reduce the bargaining power of the juice industries, establishing a market strongly oriented toward price competition.

In 2010, prices started to rise again in the first half of the year, due to the expectation of lower production in both of the world's major orange-producing states: São Paulo and Florida. However, the improvement in prices of the commodity signaled by the New York Stock Exchange was not felt immediately by the Brazilian industries, since most of the juice is sold on the European market, where price changes take longer to have an impact because sales take place through contracts with terms of several months at pre-set prices. The juice that is being delivered today may have been negotiated over 12 months ago at the current price at that time. Furthermore, European bottlers buy more juice when prices are low, scaling receipt of the juice according to their demand. Conversely, when prices rise, they begin to buy only what is required for the next three or four months, awaiting a further drop in prices. It's a battle of the giants.

The price increase in the first half of 2010 due to a reduced supply of oranges from Florida and São Paulo gave new vitality to the citrus industry, but it demonstrates a disturbing environment, marked by sharp fluctuations. This fluctuation of prices – with averages between July and June of each growing season for FCOJ prices on the New York Stock Exchange ranging from US$ 0.6895 (2003-04) to US$ 1.8083 (2006/07) per pound of soluble solids, and with average prices in the European physical market ranging from US$ 700 to US$ 2,000 per ton – reflect a sector of tremendous instability of income and high risk for anyone with capital invested in this activity.

An analysis of a 20-year period shows that – in both São Paulo and Florida – in a particular growing season the average price of orange juice could be 75% above the average price in the previous season, and on another occasion, the price could fall by nearly 45%. This oscillation occurs similarly in the price per box of oranges. The most aggravating factor for those who are in this business is the policy adopted by the international retail sector of maintaining the prices to end consumers, which can be done because juice concentrate represents a cost that is only 17% to 25% of the overall cost of the final bottled product. Lower prices in supermarkets could increase the demand for the product, but maintaining the same prices opens the way for consumers to try and/or start consuming other beverages.

So, the entire fluctuation has to be absorbed in the production and industrial links of the supply chain, as shown in Graph 31.

In the futures market, unlike the physical market, the orange juice commodity is traded exclusively on the Intercontinental Commodity Exchange (ICE), based in the United States, by means of futures and options contracts of frozen concentrated orange juice (FCOJ), US Grade A quality, at no less than 62.5° Brix. Such trading began in 1967 on the New York Cotton Exchange, the predecessor of the New York Board of Trade, which was acquired by the ICE in 2007. Options began being traded in 1985. Among the characteristics of these contracts are the physical delivery upon maturity, the months of January, March, May, July, September and November; and the size is 15,000 pounds (equivalent to 10,309 kg). The registered warehouses are located in the states of Florida, New Jersey and Delaware, and the produce that is eligible for delivery must have originated in the United States, Brazil, Mexico or Costa Rica. It is currently used by market players (growers, industries and bottlers) as a hedging tool and for FCOJ pricing, which indirectly influences the prices of NFC and other types of citrus juice as well as the price per box.

The number of outstanding contracts at the end of the 2009/10 season was equivalent to a volume of 546,000 tons of FCOJ, only 27% of worldwide production. Thus, if on the one hand the futures market is an interesting tool for hedging commodities, on the other it provides little liquidity for orange juice in relation to the volume traded globally, restricting its widespread use in the sector as a tool for risk management.

Since the 2005/06 growing season, the volume of contracts traded has fallen, from 1.4 million contracts to 942,000 in the 2009/10 season. This decrease can be explained by the numbers of speculators that have gotten out of the market due to the financial crisis, and was felt by several traded commodities. In the 2005/06 season, the equivalent volume in tons of FCOJ traded on the ICE between futures contracts and option contracts was 14.5 million, or 6.5 times the total world production. In the 2008/09 season, the equivalent of 5.8 million tons was traded, or 3.9 times overall world production. In 2009/10 there was a slight recovery in the volume traded on the stock exchange, returning to 4.8 times the worldwide production of juice, due to the more attractive prices resulting from the reduction in the size of the harvest in Florida after recent weather problems.

Price variation of 40.8 kg boxes of oranges in relation to previous years

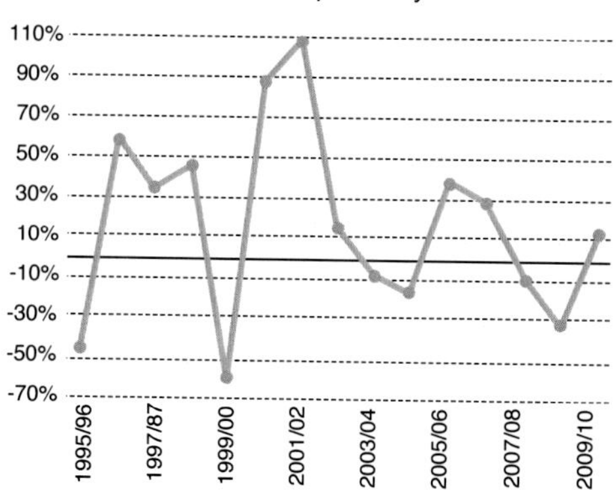

The price per box of oranges oscillated between -61% and 108%

Price variation of 40.8 kg boxes of oranges on the NY Stock Exchange in relation to previous years

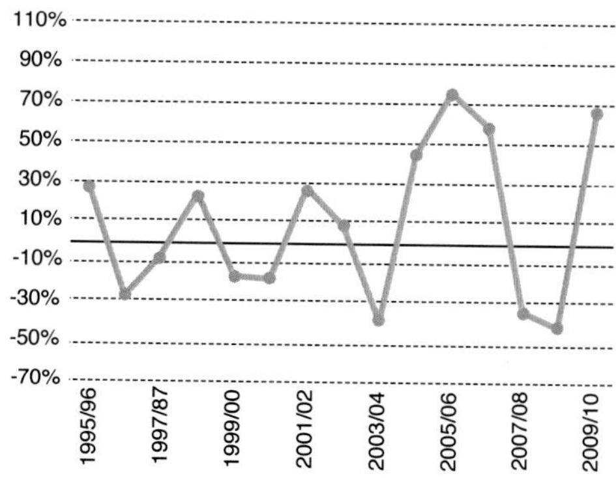

The price of orange juice oscillated between -43% and 75%

Price variation of orange juice in Germany and the USA

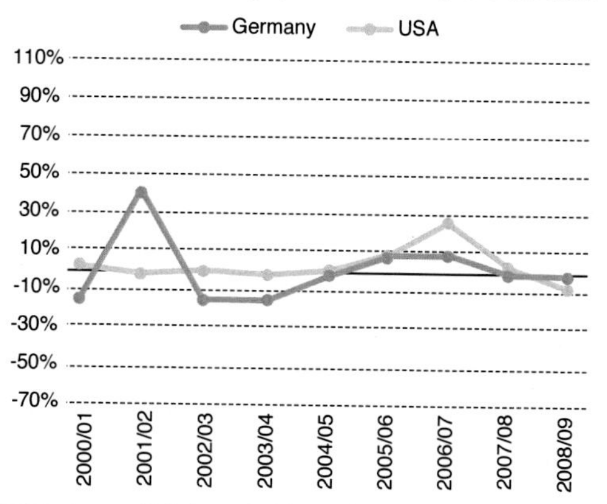

The price of reconstituted orange juice oscillated between -25% and 41%

Graph 31. **Difference in the intensity of variation in the price of oranges, FCOJ on New York stock exchange and orange juice on the retail market.**

Source: prepared by Markestrat based on the data collected from CitrusBR, CEPEA, ICE and Nielsen.

28. Breakdown of the price of orange juice on the retail market

In order to give a breakdown of the costs incurred on orange juice up to the purchase thereof by the end consumer, this topic is dedicated to the breakdown of the price starting with the final sale value as reconstituted juice on the shelves of retailers in Germany, the biggest consumer of Brazilian FCOJ, and ending with the residual value that would cover orange production costs and profit margins of growers and industries in Brazil. Some factors, such as poorly accessible data and differences in legislation, prevented a broader analysis of the European market, which accounted for 71% of Brazilian FCOJ exported in 2009.

The information for this fiscal year was gathered in interviews and contacts with Brazilian orange juice industry executives, CitrusBR, European bottlers, and their suppliers of services and inputs. Data provided by CitrusBR relating to the costs of manufacturing, warehousing, logistics, international distribution, sales, working capital financing, and revenue from FCOJ by-products for Brazilian industries were audited by a renowned international firm.

After the FCOJ has been delivered to bottlers at the port terminals in the Netherlands and Belgium, there is no further involvement by Brazilian industries, and its customers are the ones responsible for the activities of bottling and distributing reconstituted orange juice to the shelves of retailers in Germany and other European countries.

Since there is heavy competition in this stage of the supply chain, the data are treated as strategic and confidential information, and are therefore difficult to collect. More recently, retailers – realizing the importance of its sales data – have changed the way they relate to traditional research institutes. Whereas before they shared their information with well-established research institutes so that these institutes could share such data with the market, now they're treating this information as strategic, using it as leverage to increase their bargaining power in relation to the bottlers.

One example is the large retailers in Germany, such as Lidl or Aldi, which in 2009 held 18% and 15.8% shares of sales of non-alcoholic beverages, respectively; for strategic reasons, both have stopped providing research institutes with any kind of data on products sold in their retail chains. This has led to conflicting data on price and sales volumes of orange juice on store shelves in the information provided by various research institutes such as Nielsen, Canadean, Euromonitor, Eurodata, IRI and GFK, referring to European markets with significant importance in consumption, such as Germany, Spain, France and the UK.

In the United States it's no different. Data on the volume of orange juice consumption and prices on store shelves periodically published in the Citrus Reference Book and supplied by Nielsen are partial, since this institute recognizes and reports only 38% to 40% of the volume consumed in the US. For example, in the period between October 2008 and September 2009 (the most recent growing season in Florida), while the consumption of orange juice estimated by CitrusBR was 851,000 tons of 66° Brix equivalent FCOJ, Nielsen's Citrus Reference Book of August 2010 reported sales in supermarkets of 482.5 million gallons of ready-to-drink orange juice, equivalent to 341,231 tons. In that same report, US demand was reported at 1.231 million gallons, equivalent to 833,000 tons of 66° Brix FCOJ, based on sources of the reports of the processors, FASS Citrus Summary, US Census Bureau and Nielsen.

Another difficulty is that sales volumes of juices jointly with nectars are often reported to research institutes, as observed in the case of Spain and Italy, and it is therefore impossible to have a clear idea about retail prices and the quantity sold per type of beverage.

The regulations governing mandatory Brix content for juices, nectars, and still drinks vary among European countries, and there are doubts as to whether certain bottlers are complying with the minimum limits stipulated. Similarly, there is a difference in the percentage of value added tax (VAT) and the degree of formality in the business in less developed regions.

Gross margins on orange juice range from 9% to 60% depending on the sales channel, such as buying clubs, hypermarkets, supermarkets, convenience stores, bakeries, fast-food chains, restaurants, bars and hotels, according to their cost structure for distribution to the final consumer, in addition to their profit expectations.

The average distances from the orange juice port terminals in Belgium and the Netherlands to the main consumer centers range from 101 km to Brussels, to 454 km to Frankfurt, and as far as 2,714 km, in the case of Athens. This variation makes it difficult to calculate the actual amount of freight paid by bottlers for shipping FCOJ to their factories where the product is packaged.

No data are available regarding the volume of orange juice packaged per format, type or size of packaging, which can range from the traditional cartons (bricks) or more sophisticated cartons (prisms) for distribution at ambient temperature or in refrigerated systems, to PET bottles with ordinary or multilayer plastic, special labels and handles with paper or plastic labels, and aluminum bags (pouches). In addition to the different designs, they are offered in different sizes, which have volumes ranging from (for example) 400 ml to more than two liters. This complicates the collection of data regarding the incidence of packaging costs for the bottlers. Other factors that also impact cost are the design and typography of the trademark, annual purchasing volume of the bottler (which will determine the discount offered thereto), and distance from the packaging factory to the bottler.

There are several factors that must also be considered for estimating the operating costs of packaging orange juice. Among them are the loss of orange juice and of packages, which vary among filling lines of the same bottler and among different bottlers as well; type of packaging to be used; process infrastructure and level of technology adopted; utilization of installed capacity and operational efficiency of each plant; the amount of investment in equipment and facilities; capacity to finance the cycle of working capital at competitive rates; and economic factors that determine the cost structure of a country, such as interest rates, availability of credit, and foreign exchange rates. In addition, there are investments in advertising and sales promotions that do not exist for bottlers of private label retailers, but fluctuate for bottlers with their own brands.

An analysis of the balances available to the public also shows differences in reported profit or loss margins, depending on the market in which the bottler operates and the type of product packaged, whether juice, nectar or still drink.

Due to the above-mentioned difficulties, several premises were adopted for the breakdown of the price of reconstituted orange juice on the shelves of retailers in Germany.

The exchange rate used was US$ 1.30 to €1.00, the average industrial yield considered for Brazil was 238 boxes of oranges (40.8 kg ea.) for the production of one ton of 66° Brix FCOJ, a minimum Brix required by law in Germany in reconstituted juice of 11.2° Brix, the specific weight of 1.04497 kg per liter of bottled reconstituted orange juice, average cost of one-liter carton packaging for use in ambient-temperature supply chains as well as in refrigerated supply chains, and selling prices of orange juice on store shelves in Germany ranging from €0.71 to €1.05 per pound that reflect historical price levels (Table 21).

It can be concluded from the period presented in Table 22 that for a selling price of reconstituted orange juice of €0.89 per liter, equivalent to US$ 6,525 per ton of 66° Brix FCOJ or US$ 27.37 per 40.8 kg box of oranges, 28% of this amount (US$ 1,989 per ton of 66° Brix FCOJ) is paid to the orange processing industries already placed at their port terminals in Belgium and the Netherlands. When the operating costs in Brazil and abroad are deducted from this value, it can be concluded that the residual amount that would cover the cost of orange production as well as the profit margin for growers and orange juice processing industries is only 22% of the selling price on store shelves, or €0.19 per liter, equivalent to US$ 1,456 per ton of 66° Brix FCOJ or US$ 6.11 per 40.8 kg box of oranges.

One can see that 41% of the selling price of orange juice on store shelves is destined to the payment of taxes (VAT and import tariffs) and the retailer's gross margin, the equivalent

Table 21. History of monthly prices of reconstituted orange juice in Germany in euros per liter of juice diluted to 11.2° Brix.

	Harvest					
	2004/05	**2005/06**	**2006/07**	**2007/08**	**2008/09**	**2009/10**
July	0.80	0.73	0.83	1.00	0.99	0.91
August	0.71	0.74	0.84	1.00	0.97	0.88
September	0.73	0.75	0.84	1.01	0.96	0.86
October	0.77	0.75	0.84	1.02	0.96	0.88
November	0.73	0.76	0.85	1.02	0.93	0.87
December	0.73	0.76	0.84	0.99	0.97	0.86
January	0.73	0.74	0.84	1.04	0.96	0.87
February	0.72	0.74	0.84	1.05	0.97	0.86
March	0.74	0.75	0.84	1.05	0.97	0.86
April	0.71	0.74	0.85	1.04	0.96	0.85
May	0.73	0.81	0.94	1.03	0.97	0.87
June	0.75	0.84	0.99	0.97	0.96	0.85
Average	0.74	0.76	0.86	1.02	0.96	0.87

Source: prepared by Markestrat based on data from CitrusBR.

28. Breakdown of the price of orange juice on the retail market

Table 22. Breakdown of the retail price in Germany for orange juice packed in cartons of 1 liter for distribution in refrigerated chain or at room temperature.

Premises of the calculation: Euro exchange rate
Minimum Brix of reconstituted juice in Germany
Specific weight of 1 liter of orange juice reconstituted to 11.2° Brix
Average industrial yield in Brazil in the last 15 growing seasons for one ton FCOJ at 66° Brix
Price on store shelves in Germany
Expenditure on value added tax (VAT)
Price net of tax on store shelves
Retailer's gross margin
Selling price of bottler ex works retailer's distribution center
Margin of the bottler
Cost of bottler's working capital for 45 days at an interest rate of 3% per annum
Cost of shipping the bottled juice (from the bottling site to the retailer's distribution center)
Spending on advertising, marketing and sales promotions of orange juice
Cost of loss of juice in the bottling process (percentage of cost of FCOJ delivered to the bottler, including import duties)
Cost of loss of packaging materials in the bottling process (percentage of cost of packaging materials delivered to the bottler)
Cost of packaging materials (1-liter cartons, screw cap, trays, plastic film, wooden pallets, glue, outer label)
Cost of the activity of bottling the juice (manpower, utilities, maintenance, CIP, wastewater treatment)
Residual value for FCOJ delivered to the bottler's factory
Cost of bulk shipping of FCOJ (leaving the port terminals in Belgium and the Netherlands to the bottlers in Germany)
Residual value for FCOJ at the industry's terminals, including import duties
Expenditure on Brazilian FCOJ import duties for the European Union
Residual value for FCOJ at maritime terminals in Europe excluding import duties
Cost of external operations (unloading, warehousing, shipping, sales of FCOJ, administration and working capital financing)
Cost of operations in Brazil (bulk shipping of FCOJ from the factories to the Port of Santos, storage, shipping, customs clearing and Codesp fees)
Cost of operations in Brazil (industrialization, warehousing in factories, administration and marketing, working capital financing and revenue from by-products)
Residual value to cover the cost of oranges and margins of growers and industries in Brazil

Source: prepared by Markestrat based on interviews.

US$ 1.90 11,20° Brix 1,04497 kg 238 bx/t	Euros per liter of juice diluted to 11.2° Brix	US$ per ton of FCOJ at 66° Brix	US$ per 40.8 kg box of oranges	Share of the item in the value chain
	0.890	6,525	27.37	100%
19.0%	-0.142	-1,042	-4.37	16%
	0.748	5,483	23.00	
25.0%	-0.187	-1,371	-5.75	21%
	0.561	4,112	17.25	
2.0%	-0.011	-82	-0.34	1%
0.4%	-0.002	-15	-0.06	0.2%
	-0.040	-293	-1.23	4%
0.0%	0.000	0	0.00	0%
1.5%	-0.005	-35	-0.15	1%
1.0%	-0.001	-8	-0.03	0.1%
	-0.112	-821	-3.44	13%
	-0.077	-564	-2.37	9%
	0.3127	2,293	9.62	35%
379 km	-0.008	-61	-0.26	1%
	0.3044	2,232	9.36	34%
12.2%	-0.033	-243	-1.02	4%
–	0.2713	1,989	8.34	30%
	-0.022	-158	-0.66	2%
	-0.011	-79	-0.33	1%
	-0.040	-296	-1.24	5%
	0.199	1,456	6.11	22%

of US$ 2,655 per ton of FCOJ, or US$ 11.14 per box of oranges. As these three items are ad-valorem, the higher the price of reconstituted juice on store shelves, due to the higher prices of oranges and FCOJ, the greater its impact on the distribution of values throughout the chain.

It is essential to note the importance of the cost of packaging, which is essential for the orange juice to be delivered to store shelves in such a way as to preserve all of its characteristics, attributes and qualities. While packing materials cost € 0.112 per liter, equivalent to US$ 821 per ton of FCOJ, or US$ 3.44 per box of oranges, the other items that make up the costs of bottlers add up to € 0.136 per liter, or US$ 998 per ton of FCOJ, or US$ 4.19 per box of oranges, which is 15% of the shelf price.

In Table 23 we can see the breakdown of the main results of several fiscal years based on the varying prices of reconstituted orange juice on store shelves in Germany.

Table 23. Simulation based on different prices of orange juice reconstituted in Germany.

Price on store shelves in Germany + FCOJ import duties + retailer's gross margin			Value Added Tax (VAT)			Residual value for FCOJ at the industries' terminals excluding Import Duties		
Per liter of juice diluted to 11.2° Brix	Per ton of FCOJ at 66° Brix	Per 40.8 kg box of oranges	Per liter of juice diluted to 11.2° Brix	Per ton of FCOJ at 66° Brix	Per 40.8 kg box of oranges	Per liter of juice diluted to 11.2° Brix	Per ton of FCOJ at 66° Brix	Per 40.8 kg box of oranges
€ 1.090	US$ 7.991	US$ 33.52	-€ 0.449	-US$ 3.294	-US$ 13.82	€ 0.379	US$ 2.781	US$ 11.67
€ 0.990	US$ 7.258	US$ 30.44	-€ 0.406	-US$ 2.975	-US$ 12.48	€ 0.325	US$ 2.385	US$ 10.01
€ 0.890	US$ 6.525	US$ 27.37	-€ 0.362	-US$ 2.655	-US$ 11.14	€ 0.271	US$ 1.989	US$ 8.34
€ 0.790	US$ 5.792	US$ 24.29	-€ 0.319	-US$ 2.336	-US$ 8.46	€ 0.217	US$ 1.593	US$ 6.68
€ 0.690	US$ 5.058	US$ 21.22	-€ 0.275	-US$ 2.016	-US$ 8.46	€ 0.163	US$ 1.197	US$ 5.02

Source: prepared by Markestrat based on interviews.

Average cost of orange juice industrialization, from the factory gates in Brazil to the overseas port terminal in Europe		Orange Industry Europe	Residual value to cover the cost of oranges and margins of the growers and industries in Brazil			
Per liter of juice diluted to 11.2° Brix	Per ton of FCOJ at 66° Brix	Per 40.8 kg box of oranges	Per liter of juice diluted to 11.2° Brix	Per ton of FCOJ at 66° Brix	Per 40.8 kg box of oranges	Of the item in the value chain
-€ 0.073	-US$ 533	-US$ 2.24	€ 0.307	US$ 2.248	US$ 9.43	28%
-€ 0.073	-US$ 533	-US$ 2.24	€ 0.253	US$ 1.852	US$ 7.77	26%
-€ 0.073	-US$ 533	-US$ 2.24	€ 0.199	US$ 1.456	US$ 6.11	22%
-€ 0.073	-US$ 533	-US$ 2.24	€ 0.145	US$ 1.060	US$ 4.45	18%
-€ 0.073	-US$ 533	-US$ 2.24	€ 0.091	US$ 664	US$ 2.78	13%

Mapping and quantification of the citrus sector – 2008/09 harvest

The worldwide superiority and uniqueness of the Brazilian citrus sector are again recognized in a quantification study conducted in 2010, using the scientific method known as Strategic Management of Agro-Industrial Systems, or GESis (Gestão Estratégica de Sistemas Agroindustriais), developed by Professor Marcos Fava Neves, full professor of the University of São Paulo's Ribeirão Preto School of Economics, Administration and Accounting, and scientific coordinator of Markestrat (Center for Research and Projects in Marketing and Strategy).

The first quantification study carried out in 2004 was innovative in that it presented sector figures with scientific rigor. Now, with the same approach, the study delves deeper into the subject, bringing unprecedented data addressing everything from the supplies/inputs used in production to the citrus products available to consumers on store shelves throughout the world. This would not have been possible without the support of various governmental and research institutions whose studies have been contributing to Brazil's worldwide leadership in citrus production. In order for the study to be completed, it was also fundamental for data to be made more openly available on the part of agricultural supply companies, growers, juice exporting industries, smaller-sized orange juice factories, packing houses, bottling plants and distribution channels.

With the information collected, estimates were made of sales and financial operations in the sector for the 2008/09 season.

Here we probably have the most up-to-date picture of the productive chain in Brazil. This material serves as a stimulus for public and private decision-making, because it shows the strong interconnection between the links in the production chain and their ability to generate revenues, taxes and jobs.

In this study, the GDP (Gross Domestic Product) of the citrus sector was calculated for the 2008/09 agricultural year, estimated at US$ 6.5 billion (Table 24), around 2% of the GDP of Brazilian Agribusiness, of which US$ 4.39 billion is generated domestically and US$ 2.15 billion internationally. Of the citrus sector's GDP, 34% is from sales of oranges (fresh fruit) on the domestic market and 28% is from juice exports (FCOJ and NFC). It is noteworthy that juices account for 94% of the exported value. By dividing citrus sector's GDP by the area of citrus croplands in Brazil (according to the IBGE), an amount of R$ 6,700 per hectare is obtained, twice the GDP of the sugarcane sector per hectare of cropland (R$ 3,300). The calculation of the citrus sector's GDP was estimated by taking the sum of sales of final goods in the citrus agribusiness system.

Figure 7 represents the citrus agro-industrial system and the values below each link indicate the gross sales of this segment with the citrus sector in the 2008/09 season. The total gross revenue of the citrus sector this year was roughly US$ 14.6 billion. This value represents the sum of estimated sales in the various links of the citrus production chain and financial operations of facilitating agents.

Table 24. Estimate of gross domestic product of the citrus sector based on final products.

Product	Domestic market (DM) US$ (millions)	Foreign market (FM) US$ (millions)	Total (DM, FM) US$ (millions)
Orange	2,232.9	19.1	2,252.0
Lime	673.1	48.2	721.2
Tangerine	945.9	5.8	951.7
FCOJ	-	1,545.9	1,545.9
NFC	-	299.5	299.5
Citrus pulp pellet	85.2	93.5	178.8
Essential oils	-	72.9	72.9
Terpene	-	55.2	55.2
Frozen cells	-	9.1	9.1
D-limonene	-	0.9	0.9
Orange juice/nectar	459.1	-	459.1
Total	4,396.21	2,150.10	6,546.31

Source: Neves and Trombin based on data collected by Markestrat (2010).

Before the farms

The industry of agricultural supplies earned US$ 819 million from sales to the citrus sector in the 2008/09 season. Graph 32 summarizes all sales in this link of the chain.

Sales from the fertilizer industry to the citrus sector totaled US$ 210.1 million, and from the pesticide industry, US$ 288.2 million. Sales of pesticides rose 75% in relation to 2004. Due to the strict control of pests and diseases, sales of acaricides, fungicides and insecticides account for 84% of revenues. The appearance of citrus greening in Brazilian orange groves increased spending on phytosanitary control, especially in the use of insecticides and acaricides. From 2003 to 2008, demand for insecticides in the citrus sector grew from 593 to 4,060 tons of active ingredient. In the same period, demand for acaricides increased from 8,876 to 13,798 tons of active ingredient. This increase is also justified by the higher density of trees planted in the orange groves and a more favorable ratio. In 2001, 75 boxes of oranges (40.8 kg ea.) were required to acquire one ton of pesticide. In 2008, this quantity was reduced to 56 boxes.

The phytosanitary problems that affect Brazilian citrus farming make the production of seedlings an important step in the chain, because it must take place indoors and comply with state laws in force. Of the total revenue from agricultural supplies in the 2008/09 growing season, seedlings accounted for nearly 4%, or US$ 39.5 million. The use of better quality seedlings has also contributed to increased productivity in the sector.

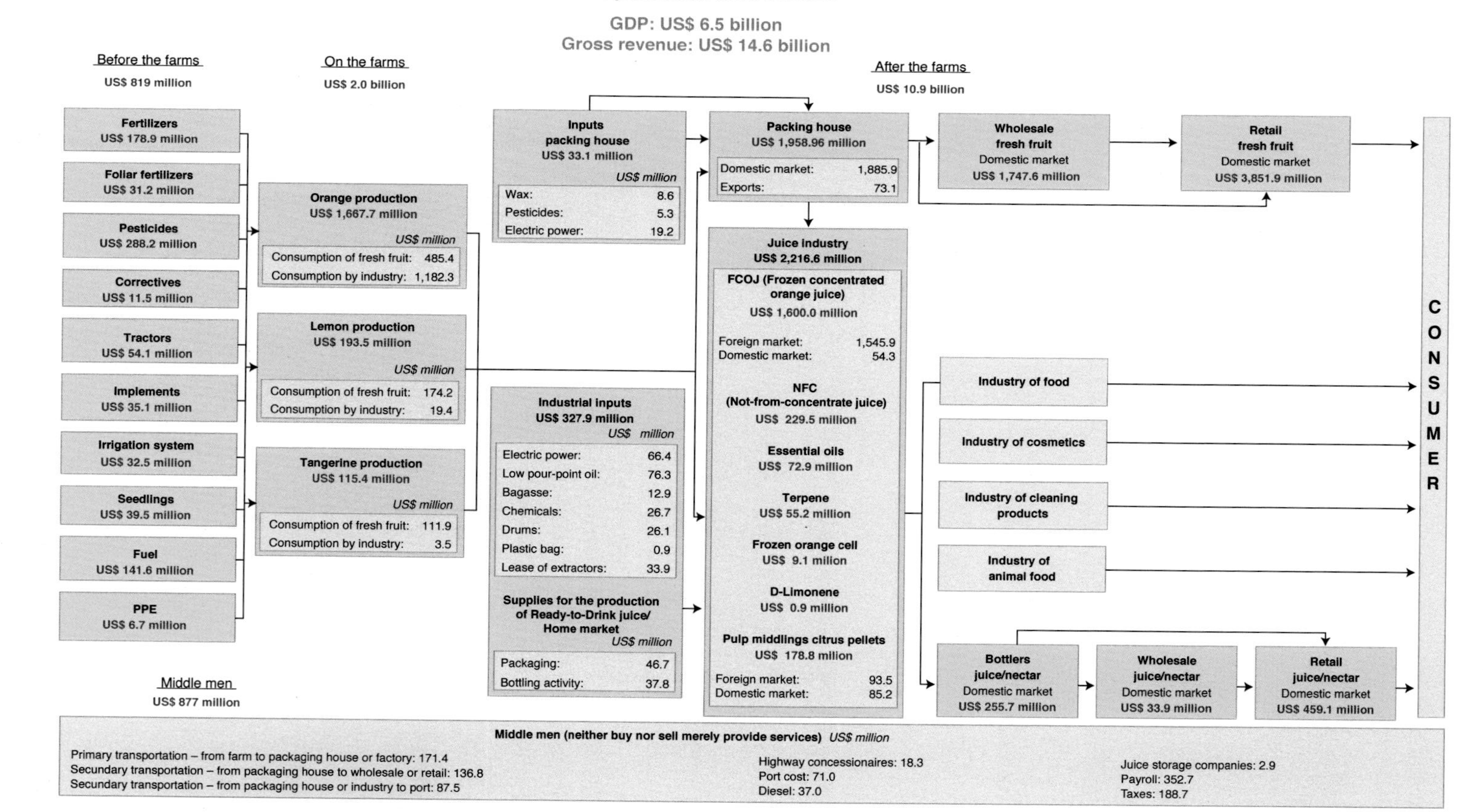

Figure 7. Citrus production chain in Brazil.

Source: Neves and Trombin, based on data from Markestrat (2010).

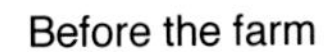

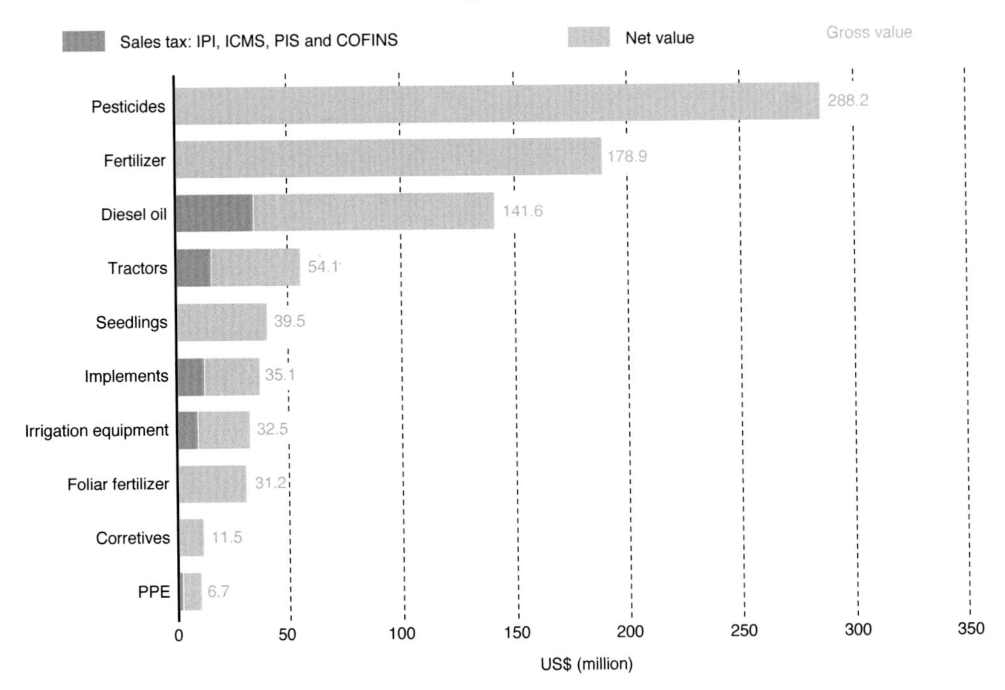

Graph 32. **Sales of agricultural supplies link.**

Source: Neves and Trombin, based on data from Markestrat (2010).

Sales of diesel fuel in the citrus sector for the 2008/09 growing season were estimated at US$ 141.6 million, and sales of agricultural implements totaled US$ 35.1 million. Sales from the tractor industry totaled US$ 54.1 million, or 1,227 units, 91% of which are in the category of tractors with engines from 50 hp to 99 hp.

On the farms

As shown in Graph 33, in the 2008/09 season, revenue from Brazilian production of citrus fruits (oranges, lemons/limes and tangerines) reached about US$ 2 billion. Of this production, around 67% was destined to industrial processing, 32% was consumed on the domestic market as fresh fruit, and 1% was earmarked for fresh fruit exports.

The average price paid to growers per box of oranges for consumption as fresh fruit was R$ 10.16. Of the oranges destined for industrial production, 35% was grown on industry-owned croplands; 34% was purchased from growers with pre-established delivery contracts at an average price of R$ 10.30/box; and 31% was purchased from growers on the spot market at an average price of R$ 7.10/box.

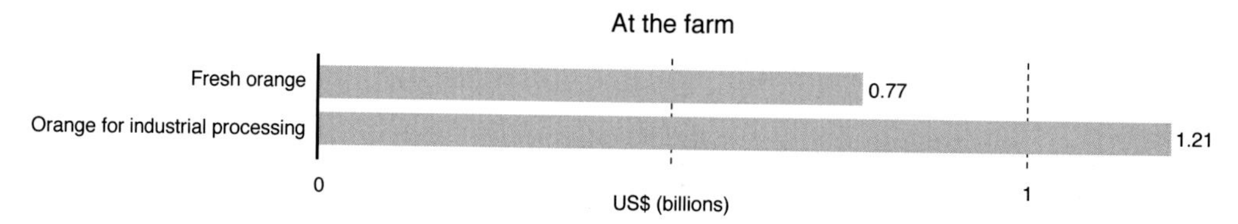

Graph 33. **Revenue from sale of citrus produce (oranges, lemons/limes and tangerines).**
Source: Neves and Trombin, based on data generated by Markestrat 2010).

After the farms

The inputs/supplies acquired by packing houses and citrus juice processing factories totaled US$ 360.9 million, as shown in Graph 34. Of this total, electricity represents 24% and low pour-point oil/Bagasse, 25%.

In the 2008/09 growing season, the revenue generated by the packing houses from fresh fruit was US$ 1.8 billion, 96% of which was on the domestic market. On the wholesale market, sales were US$ 1.7 billion. In retail, sales totaled US$ 3.8 billion, 58% of which was from the sale of oranges, 17% from lemons/limes, and 25% from tangerines.

Sales of juices and by-products totaled US$ 2.2 billion, 95% of which was on the foreign market and 5% on the domestic market. Of the total revenues from exports (US$ 2.07 billion), 86% corresponds to juices. Revenues of the bottling companies, the wholesale market and the retail market from orange juices and nectars totaled (respectively) US$ 255.7 million, US$ 33.9 million and US$ 459.1 million.

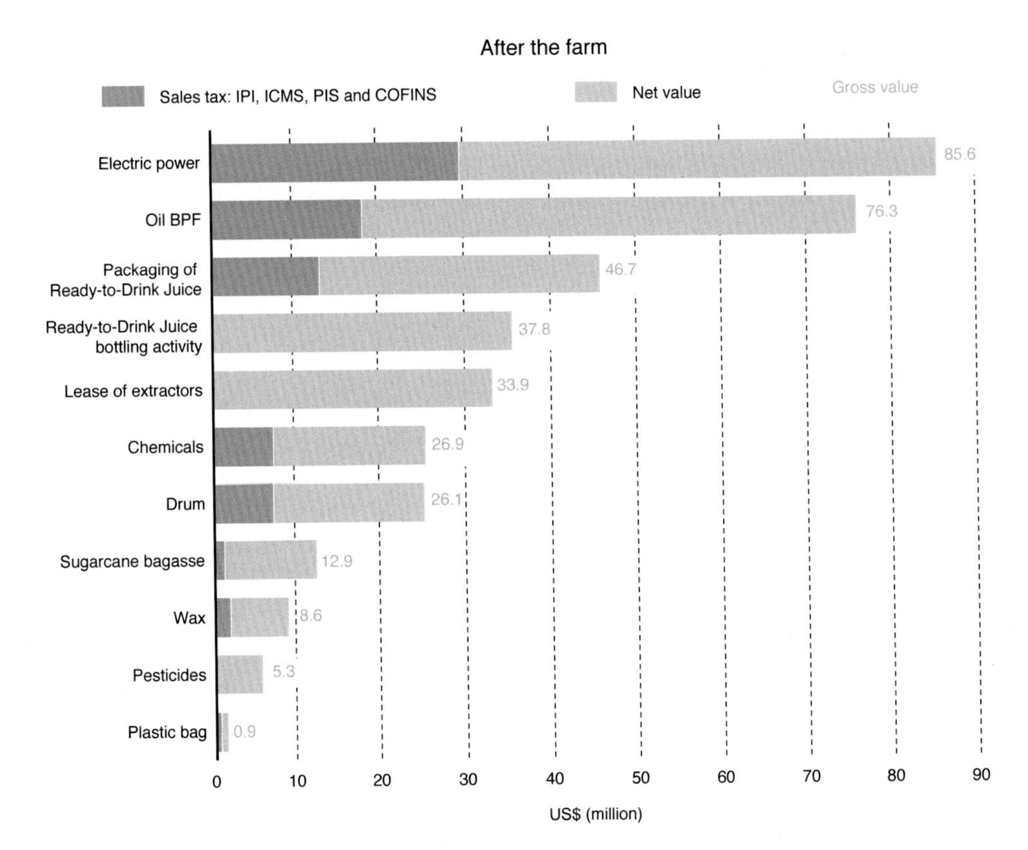

Graph 34. Sales of the industrial inputs link of the supply chain.

Source: Neves and Trombin, based on data generated by Markestrat (2010).

Facilitating agents

Revenues of the facilitating agents from the citrus production chain in the 2008/09 growing season was US$ 877.5 million, as detailed below:

- **Shipping** – In the 2008/09 season, more than six trucks an hour (on average) passed through toll booths carrying orange juice to the Port of Santos. Revenues of toll road concessionaires from the citrus production chain totaled US$ 18.3 million, or 5% of US$ 396 million that the sector spent on transportation. Diesel fuel accounted for 9% of the total. The shipping segments can be divided into primary and secondary. The primary segment refers to transportation of the fruit from the farms to the packing house or to the juice industries; the overall revenue of this segment was US$ 171.4 million (43% of the total).
- The secondary segment refers to shipping from the packing house to the wholesale and/ or retail market, with revenues of around US$ 137 million; from the packing house to the

port, with revenues totaling US$ 2.7 million; or from the factory to the port, with total revenues of US$ 85 million.

- **Port Cost (Port of Santos)** – It is estimated that, in 2008, receipts for the Port of Santos from customs clearance, elevation and supervision of loading the orange juice was on the order of US$ 71 million. It is noteworthy that 97% of Brazil's total juice exports were shipped through the Port of Santos.
- **Payroll** – The 2008/09 crop year ended with 132,776 workers in the sector: 121,332 in citrus farming and 11,444 in the juice industry. Almost 69,000 workers were hired during the 2008/09 growing season. The average monthly wages of workers in citrus farming activities was US$ 364, while the average wages of workers in the juice industry worker was US$ 864. The overall payroll in the 2008/09 growing season was US$ 352.7 million. This means that the farming of citrus fruits accounted for 91% of the jobs in the sector, and industry accounted for 9%.

Added taxes

To calculate the total amount of taxes, we used the sum of the taxes generated at each link along the production chain, from the sale of agricultural and industrial supplies to the sale of the final products. To eliminate double counting and taking into account the taxes added in the production chain, the taxes generated in the first links of the chain (agricultural and industrial inputs) were subtracted from this total. As a premise for the estimation of added taxes in the production chain, it was assumed that the companies opted for the system of taxation based on actual profit.

The results of this estimate showed that the total taxes on sales in the production chain in the 2008/09 growing season totaled about US$ 339.4 million, of which US$ 150.67 million was generated by the sale of agricultural and industrial inputs/supplies. Thus, the added taxes in the production chain were estimated at US$ 188.74 million.

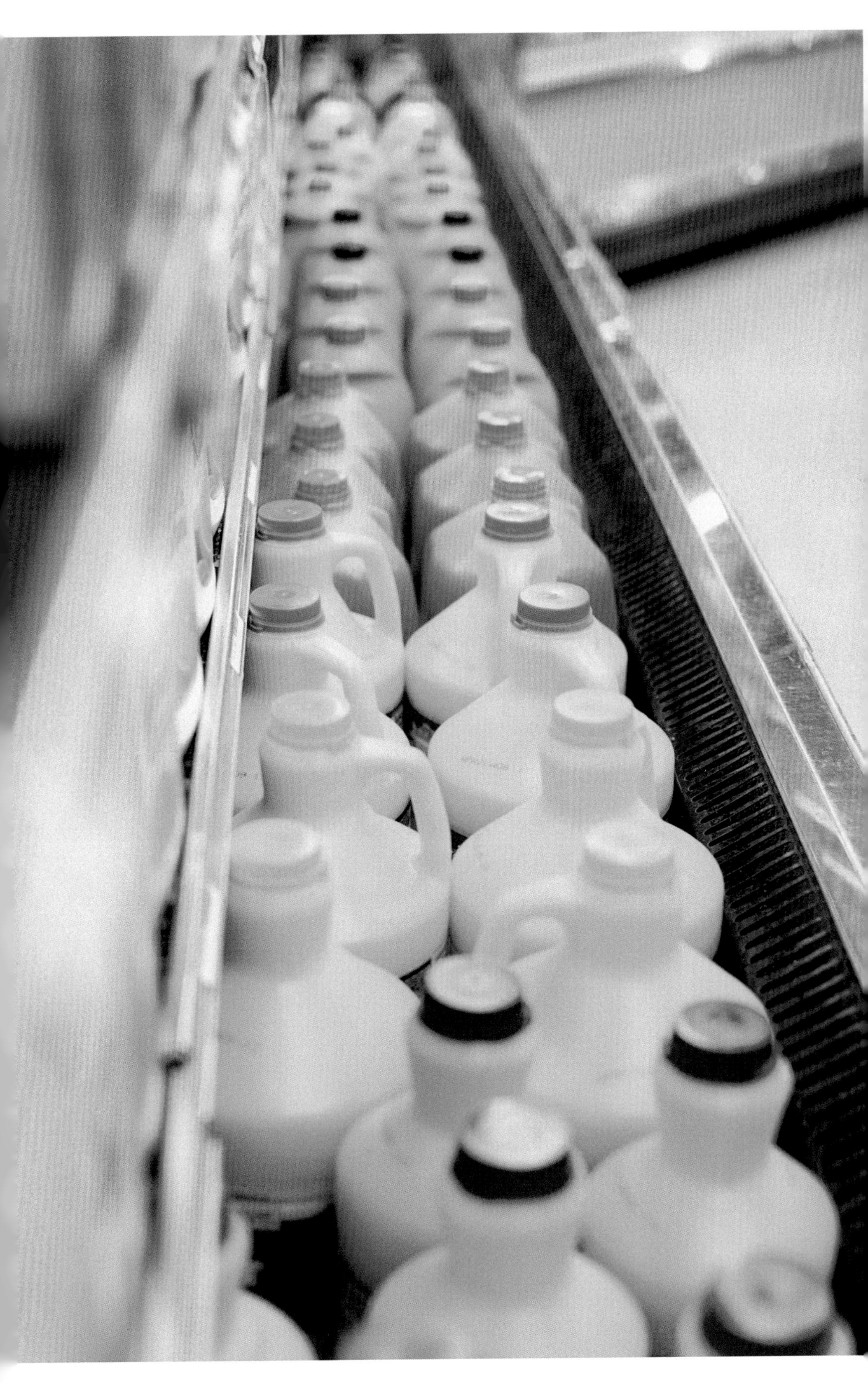

Mapping the consumption of citrus products

29. Nutritional benefits of oranges

The benefits of including fruit in one's diet are innumerable, due to its high nutritional value and high levels of fiber, water and vitamins. Oranges are no different. Consuming one unit of this fruit corresponds to the recommended daily amount of vitamin C (60 mg). This powerful vitamin increases protection against infections and has healing properties as well as high antioxidant protection power. Antioxidants protect the body from the harmful action of free radicals. One would have to eat 15 apples to get the same amount of vitamin C found in one orange.

Oranges also facilitate intestinal regularity because of the high soluble fiber content, found in the pulp and bagasse. The white part of the bagasse also contains pectin, which prevents cancer and helps lower cholesterol in the body.

The amount of calcium in oranges helps maintain bone structure and proper muscle and blood formation. Beta carotene, a phytonutrient that gives oranges their color, prevents cancer and heart attacks. Drinking orange juice daily can also help increase good cholesterol (HDL) and lower bad cholesterol (LDL). Moreover, its antioxidants improve the functioning of blood vessels, helping to prevent several forms of heart disease.

The "5-a-day" campaign is well known in Europe, promoting healthy eating habits that include eating five servings of fruits and/or vegetables daily. A glass of fruit juice is regarded as one serving.

30. Definition of juice, nectar and still drink

Although not widely known to consumers in general, the difference between juice, nectar and still drink is related to the content of fruit juice present in the packaged beverage. Worldwide, products labeled as "juice" must contain 100% fresh fruit, therefore these are pure products with no preservatives or sweeteners and no artificial colors, and may or may not contain pulp of the fruit itself. In this category, there is a division between "Reconstituted Juices," which are basically concentrated from three to six times at the juice concentrate factories where they are produced, and subsequently diluted with potable drinking water at a bottling plant, returning the juice to its original condition (in terms of concentration of soluble solids in water) at the time of bottling, before being distributed to consumers. Another division of the juice category is "Not-From-Concentrate," commonly known as NFC, which only undergoes a slight pasteurization process.

In the nectar category, the packaged beverage has a smaller content of pure juice, ranging from 25% to 99% depending on the laws of each region around the world. Unlike juice (100% juice), nectar can contain sweeteners, coloring and preservatives – additives that are generally cheaper than the soluble solids of the fruit itself, making this category more affordable for consumers in the intermediate per-capita income range.

In the still drink category, the juice content in the packaged beverage is less than 25%, and in many countries only 3% to 5% (for example, China). These beverages contain a larger quantity of additives, making them a product of lesser value, representing a gateway for the consumption of industrialized still drinks for lower income populations.

31. World consumption of beverages

Over the past seven years, the world population has increased at a rate of 1.2% per year, and the consumption of commercial beverages has grown 3.6% per year. Therefore, during this period a market of 297 billion liters of beverages was created, bringing the total market for commercial beverages to approximately 1.6 trillion liters in 2009, equivalent to 231 liters per capita per year. The leading beverage category in market share in 2009 was hot tea (at 20.9%), followed by bottled water (15.3%), milk (12.8%), carbonated soft drinks (12.5%), beer (11.2%), hot coffee (8.2%), still drinks (2.7%) and juices/nectars (2.6%), as shown in Graph 35.

With a 35% share in the segments of juices and nectars, orange flavor corresponded to 0.91% of the worldwide beverage market in 2009. In the segment of still drinks, with a 30% share, orange flavor corresponded to 0.82% of the total.

From 2003 to 2009, the fastest- growing beverage categories, in terms of consumption, were those of lower added value and low calorie content, as shown in Graph 36. Fruit-flavored still drinks increased by 7.3% per year; bottled water by 6.6% per year; milk-based beverages, 6.5%; and hot teas, 4%. Juices and nectars grew by 2.1%.

Consumers around the globe are increasingly aware of price, but also follow the trends of health, well-being, responsible consumption, and convenience. This price-oriented behavior increased after the 2008 crisis, when consumers began to worry more about financial planning, reevaluating the need to buy fancy products and starting to appreciate sale promotions.

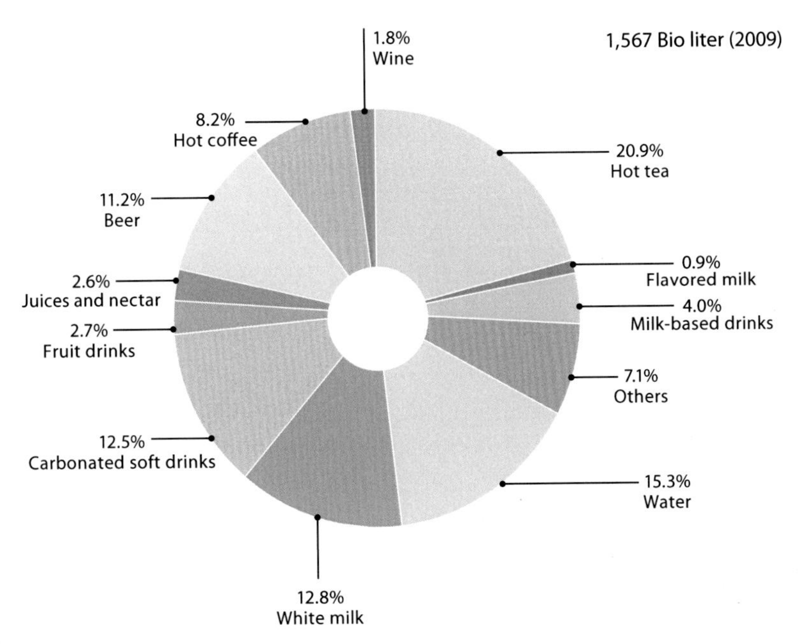

Graph 35. Share of the world market, by beverage category.

Source: prepared by Markestrat based on data from Euromonitor.

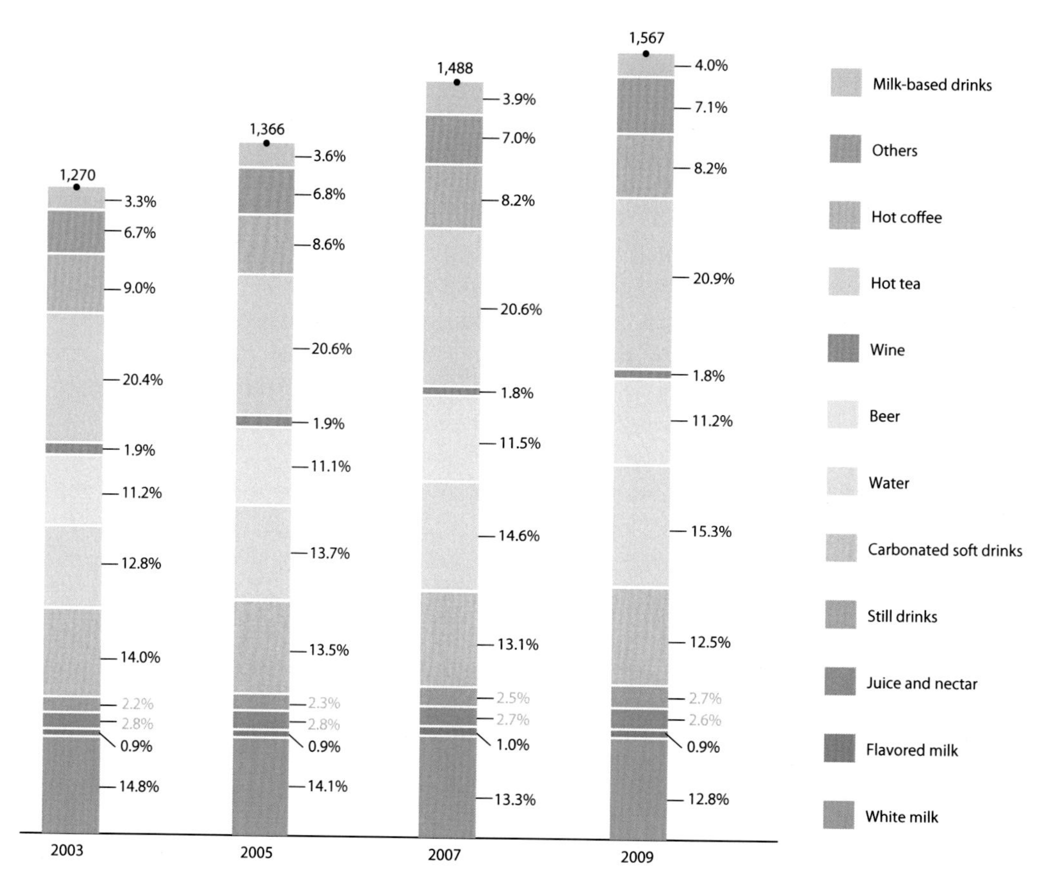

Graph 36. **Evolution of market share per type of beverage.**

Source: prepared by Markestrat based on data from Tetra Pak and Euromonitor International.

32. World consumption of fruit juices, nectars, and still drinks

In 2009, the world consumed 117.7 billion gallons of industrialized still drinks. Of the total volume, 77% were consumed in 40 countries, with 23.5 million liters in the juice category, 17 million in the nectar category, 42 million in the category of still drinks, and 35 million in the category of powdered and concentrated juices. In the period from 2003 to 2009, the consumed volume of fruit-based beverages increased by 30.2%. However, since much of this growth came from increased consumption in lower social classes in emerging countries, the increase in the sales volume occurred primarily in the categories of nectars and still drinks, and therefore does not reflect an increased demand for orange juice at 66° Brix, because these are categories of beverages that are diluted in water instead of 100% juice (Graph 37).

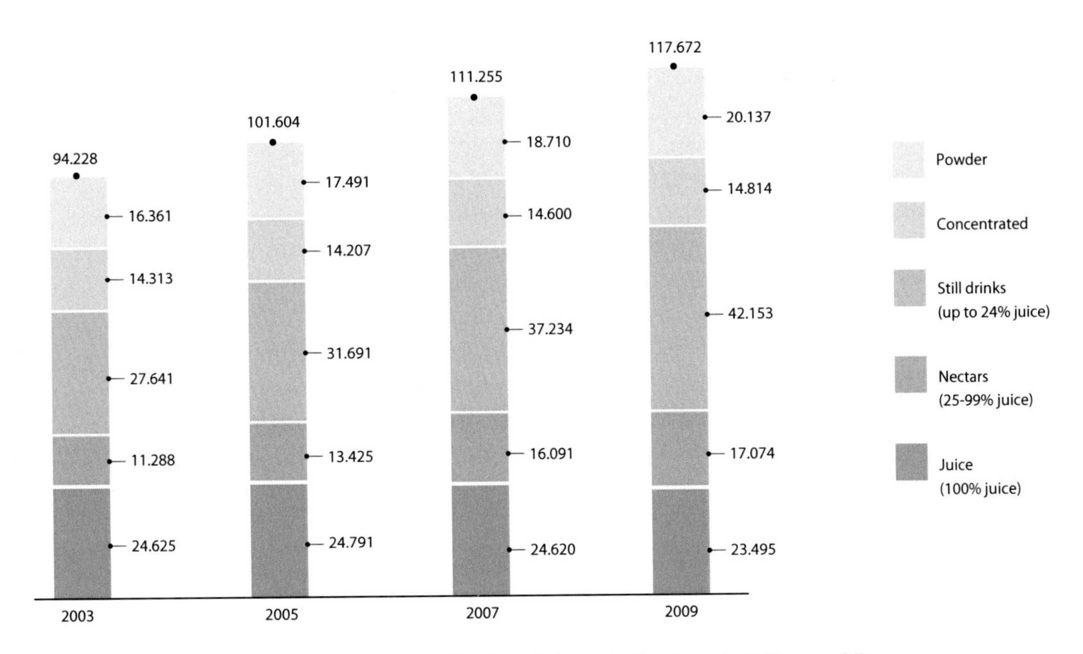

Graph 37. **Evolution of the volume of industrialized fruit drinks in billions of liters.**
Source: prepared by Markestrat based on data from Tetra Pak and Euromonitor International.

The 7% growth in sales of still drinks per year was higher than the rise in the category of industrialized beverages, at 3.6% per year, increasing the combined market share by 3.8%. Similar expansion, albeit less intense (2.2% growth per year), also occurred in the category of powdered and concentrated juices, beverages to be diluted before consumption at home, the demand for which is also higher in developing countries, particularly India.

In contrast to this growth, the juice category on the global market has shown a 0.8% decrease in consumption per year and a loss of 0.4% market share for the remaining categories, especially in the traditional markets – the United States and Europe.

In this respect, it is important to stress the relevance of emerging markets in sustaining the annual growth rate of 2.7% of ready-to-drink juices over the last seven years in these markets. In the period from 2003 to 2006, demand for ready-to-drink juices in Asia, the Middle East and Latin America grew at an annual rate of 5.9%, 4% and 2.8%, respectively. During this period, the nectar category experienced a higher rate of expansion in the 40 selected countries, at 7.6% a year, compared to 3.6% per year for still drinks and a decrease of 0.3% per year for juices. More recently, from 2006 to 2009, the growth rate of ready-to-drink beverages in these emerging markets intensified, reaching an annual rate of 9.8% in Asia, 4.6% in the Middle East, and 6.1% in Latin America. During this period, the annual growth rate of still drinks increased to 6.4% per year, the annual growth rate of nectar dropped to 2.5% per year, and the decreasing demand for 100% juices to 2.1% per year.

33. Orange flavor

Orange flavor stands out as the most widely consumed products among fruit-based beverages ready for consumption. In 2009, orange flavor had a 35% share, ahead of apple flavor, which had a 16% share (Graph 38). However, in some markets, such as the United States, consumption of apple juice has been gaining ground over orange juice consumption. In countries such as Russia, Germany, Ukraine and Turkey, apple flavor prevails over orange flavor in the juice category.

In the 40 selected countries, representing 99% of global consumption of orange flavor, a detailed analysis shows that out of the 63.5 billion liters of ready-to-drink fruit-based beverages consumed, 20.4 billion were orange flavor and 7.5 billion were apple flavor. However, in the period from 2003 to 2009, in the juice category, there was a greater diversification of flavors

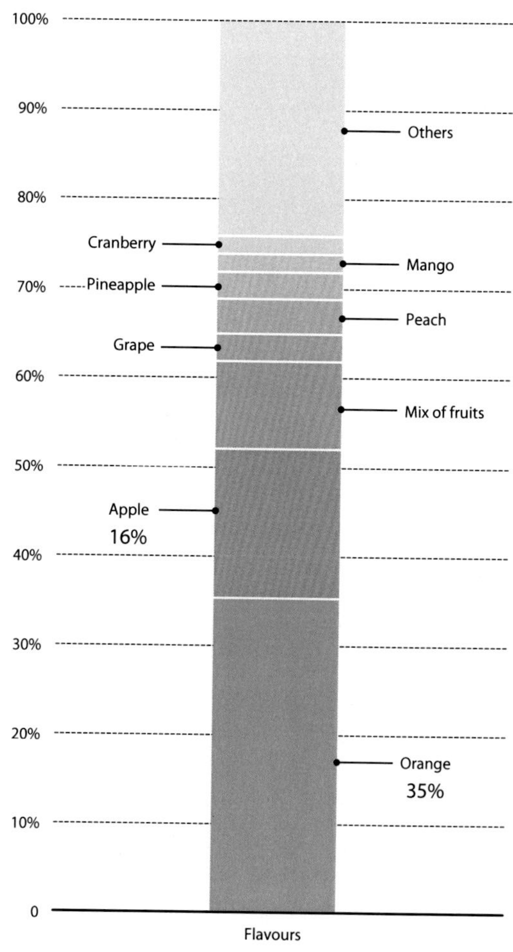

Graph 38. Share of fruit flavors in juices and nectars consumed in 2009.

Source: prepared by Markestrat based on data from Tetra Pak and Euromonitor International.

consumed, with a reduction in annual demand for orange and apple flavors of 1.6% and 2.3%, respectively, and an increase in demand for tomato flavor and multifruit flavors of 2.6% and 1.3%, respectively. In the case of nectars and still drinks, the volume of orange flavor increased, but to a lesser extent when compared to peach, grape, mango and multifruit flavors.

This diversification in the flavors consumed and the consequent loss of market share for orange flavor have contributed to the reduction in global demand for orange juice, which fell by 6% between 2003 and 2009 (Table 25). This behavior is not what one would expect considering the main demographic data of the 40 countries that together represent 99% of the world demand for orange flavor. Unlike orange juice consumption, which fell 6%, demographic indexes showed growth: the population increased 5%, overall GDP increased 51%, GDP per capita rose 43%, and per capita net income rose 40% (Table 26 and 27a, b).

The type of beverage in which orange flavor is demanded has also undergone changes over the past seven years. While there was a decrease of 1.6% in the consumption of orange juice, there was a 4% and 1.6% increase in consumption of nectars and still drinks, respectively (Graph 39). The consumption profile in the selected countries by type of beverage is related to availability of income per capita.

Regions with high per capita income, such as the United States and Europe, tend to consume 100% pure orange juices, which are more expensive due to their higher fruit content. Whereas countries with less available per capita income, such as the BRIC group (Brazil, Russia, India and China) plus Mexico, tend to consume a greater quantity of nectars and still drinks, which are more accessible beverages given their lower concentration of juice (Graph 40).

Table 25. Consumption of orange juice in 40 selected countries, converted to thousands of tons at 66° Brix FCOJ equivalent.

	2003	2004	2005	2006	2007	2008	2009	2010	Variation in the period
Total	2,415	2,414	2,392	2,361	2,309	2,245	2,275	2,288	-5%
By country									
United States	1,002	1,029	985	924	882	826	851	809	-19%
Germany	256	231	211	213	201	199	193	198	-23%
France	152	147	153	161	165	166	171	175	15%
United Kingdom	140	136	136	138	137	140	136	137	-2%
Canada	115	117	133	126	120	114	117	121	5%
China	44	42	48	56	60	68	74	88	99%
Russia	51	59	63	74	79	78	73	84	64%
Japan	92	97	95	95	92	76	74	75	-18%
Spain	43	45	47	46	46	47	47	48	12%
Brazil	45	37	40	41	37	38	41	45	0%
Poland	40	41	40	38	36	37	39	41	3%
Australia	38	40	40	40	40	39	40	40	7%
South Korea	45	43	42	40	39	39	38	39	-14%
The Netherlands	36	37	35	35	32	32	32	32	-11%
Mexico	30	29	29	30	32	30	32	31	5%
Italy	33	33	33	31	30	29	29	29	-12%
South Africa	23	23	25	26	28	27	27	27	19%
Saudi Arabia	15	16	17	19	21	22	23	26	76%
Sweden	26	24	24	25	25	24	24	24	-8%
Belgium	22	22	22	24	23	23	23	23	4%
India	19	17	16	16	17	18	19	19	-1%
Argentina	4	5	5	6	9	11	13	17	358%
Austria	19	18	18	17	17	17	17	17	-14%
Norway	12	13	14	15	15	17	17	17	42%
Switzerland	15	14	14	14	14	14	14	14	-9%
Ireland	13	13	14	14	14	14	12	12	-6%
Chile	6	6	7	8	9	11	11	11	104%
Denmark	12	12	12	12	11	11	11	10	-12%
Finland	16	13	14	13	11	10	11	10	-36%
Greece	12	11	12	12	11	11	11	10	-12%
Ukraine	5	8	11	12	14	15	11	10	87%
Indonesia	2	3	3	3	4	4	6	8	213%
Romania	3	3	4	5	6	7	7	7	163%
New Zealand	7	6	6	7	7	7	7	7	0%
Morocco	1	1	2	2	3	4	4	6	743%

Table 25. Continued.

	2003	2004	2005	2006	2007	2008	2009	2010	Variation in the period
Total	2,415	2,414	2,392	2,361	2,309	2,245	2,275	2,288	-5%
By country (continued)									
Taiwan	7	6	6	6	6	6	6	6	-13%
Turkey	3	4	5	7	7	6	5	4	62%
Israel	5	5	5	4	4	4	4	4	-20%
Philippines	3	3	3	3	3	3	3	3	0%
Colombia	4	3	3	3	3	4	3	3	-13%
By continent									
North America	1,117	1,147	1,118	1,050	1,002	939	968	930	-15%
Europe	910	886	882	906	895	895	883	903	-1%
Western Europe	797	759	748	758	743	741	738	746	-7%
Eastern Europe	113	127	134	148	153	154	145	157	38%
Asia	232	232	235	242	245	241	247	268	15%
South and Central America	88	80	84	88	90	93	100	108	16%
Oceania	45	46	46	47	46	46	47	47	6%
Africa	23	24	27	28	31	31	32	33	40%

FCOJ consumption equivalent to 66° Brix; does not include orange juice used to produce carbonated soft drinks: estimated 70,000 tons of FCOJ annually.

Source: prepared by Markestrat based on data from CitrusBR.

Table 26. Summary of key demographics in 40 selected countries.

Summary of data on the 40 selected markets		2003	2010	Variation
Population on January 1st	Inhabitants (×1000)	4,388,933	4,691,268	7%
Total GDP	Billions of US$	34,711,852	56,781,435	64%
GDP per capita	US$ per inhabitant	7,927	12,104	53%
Net income per capita	US$ per inhabitant	5,235	7,805	49%
Unemployment rate	%	8.5%	8.3%	-3%
Consumption of orange juice FCOJ equivalent to 66° Brix	Tons	2,415	2,288	-5%

Source: prepared by Markestrat based on Tetra Pak and Euromonitor International, the World Bank, and CitrusBR.

Table 27a. Relationship among demographic data and consumption of orange-flavor drinks,

	Population on January 1st (Inhabitants × 1000)				Total GDP-current prices converted at annual exchange rate (US$ billion)		
	2003	2010	Variation	Share '10	2003	2010	Variation
World	6,311,539	6,830,769	8%	100,0%			
Selected markets	4,379,029	4,691,268	7%	68,7%	34,711,852	56,781,435	64%
United States	290,211	308,862	6%	4,5%	11,142,200	14,657,800	32%
Germany	82,537	81,722	-1%	1,2%	2,442,753	3,309,241	35%
France	60,067	62,772	5%	0,9%	1,800,402	2,579,032	43%
United Kingdom	59,438	61,966	4%	0,9%	1,860,893	2,245,379	21%
Canada	31,676	33,968	7%	0,5%	866,920	1,574,533	82%
Japan	127,694	127,363	0%	1,9%	4,229,091	5,460,244	29%
Russia	144,964	141,786	-2%	2,1%	431,488	1,479,842	243%
China	1,284,530	1,341,414	4%	19,6%	1,647,918	5,959,523	262%
Spain	41,663	45,928	10%	0,7%	883,863	1,407,223	59%
Brazil	181,537	193,253	6%	2,8%	552,384	2,089,018	278%
Mexico	101,999	108,997	7%	1,6%	700,324	1,040,210	49%
Australia	20,012	22,226	11%	0,3%	540,407	1,236,720	129%
South Korea	47,860	48,910	2%	0,7%	643,760	1,014,762	58%
Poland	38,219	38,167	0%	0,6%	216,801	468,630	116%
Netherlands	16,193	16,439	2%	0,2%	538,432	783,312	45%
Italy	57,321	60,401	5%	0,9%	1,507,505	2,051,147	36%
South Africa	46,848	49,912	7%	0,7%	168,219	363,734	116%
Saudi Arabia	22,496	26,246	17%	0,4%	214,573	443,099	107%
Sweden	8,941	9,341	4%	0,1%	314,713	457,994	46%
Belgium	10,356	10,827	5%	0,2%	311,261	466,593	50%
India	1,069,041	1,215,939	14%	17,8%	591,332	1,652,787	180%
Norway	4,552	4,858	7%	0,1%	225,116	414,495	84%
Austria	8,102	8,409	4%	0,1%	252,090	376,114	49%
Switzerland	7,314	7,785	6%	0,1%	325,052	523,813	61%
Argentina	38,024	40,666	7%	0,6%	129,596	368,736	185%
Ireland	3,964	4,456	12%	0,1%	157,781	203,866	29%
Ukraine	47,824	45,783	-4%	0,7%	50,133	137,894	175%
Greece	11,006	11,290	3%	0,2%	194,661	304,825	57%
Denmark	5,383	5,534	3%	0,1%	212,968	310,405	46%
Chile	15,955	17,190	8%	0,3%	73,990	203,423	175%
Finland	5,206	5,351	3%	0,1%	164,163	238,771	45%
New Zealand	4,028	4,310	7%	0,1%	73,098	141,319	93%
Romania	21,773	21,369	-2%	0,3%	59,466	161,629	172%
Indonesia	213,655	232,517	9%	3,4%	234,665	706,834	201%
Taiwan	22,521	23,105	3%	0,3%	310,764	430,190	38%
Turkey	66,333	72,474	9%	1,1%	303,008	733,379	142%
Israel	6,690	7,518	12%	0,1%	118,903	217,952	83%
Philippines	81,534	93,923	15%	1,4%	79,634	188,719	137%
Colombia	41,741	46,300	11%	0,7%	91,703	287,134	213%
Morroco	29,821	31,992	7%	0,5%	49,823	91,117	83%

Consumption of 66° Brix equivalent FCOJ does not include orange juice used to produce carbonated soft drinks: estimated 70,000 tons of FCOJ annually,

DP per capita-current prices onverted at annual exchange ate JS$/inhabitant)			Per capita net income-current prices converted at annual exchange rate (US$/inhabitant)			Unemployment rate (%)		Consumption of orange juice (Thousand tons of FCOJ 66° Brix equivalent)			
003	2010	Variation	2003	2010	Variation	2003	2010	2003	2010	Variation	Particip, '10
			3,897	5,804	41%						
7,927	12,104	53%	5,247	7,805	49%	8.5%	8.3%	2,415	2,288	-5%	100%
38,393	47,457	24%	27,748	33,904	22%	6.0%	9.6%	1,002	809	-19%	35%
29,596	40,494	37%	20,085	26,228	31%	9.8%	6.8%	256	198	-23%	9%
29,973	41,086	37%	19,596	28,035	43%	9.0%	9.8%	152	175	15%	8%
31,308	36,236	16%	20,232	22,620	12%	5.0%	7.8%	140	137	-2%	6%
27,368	46,354	69%	16,081	27,802	73%	7.6%	8.0%	115	121	5%	5%
33,119	42,872	29%	20,814	27,350	31%	5.3%	5.1%	92	75	-18%	3%
2,977	10,437	251%	1,721	5,875	241%	8.0%	7.5%	51	84	64%	4%
1,283	4,443	246%	697	2,366	239%	4.3%	4.1%	44	88	99%	4%
21,215	30,640	44%	13,660	21,072	54%	11.5%	20.1%	43	48	12%	2%
3,043	10,810	255%	1,963	6,698	241%	9.7%	6.7%	45	45	0%	2%
6,866	9,543	39%	4,744	6,507	37%	3.0%	5.4%	30	31	5%	1%
27,004	55,643	106%	16,802	33,767	101%	5.9%	5.2%	38	40	7%	2%
13,451	20,748	54%	8,800	12,945	47%	3.6%	3.7%	45	39	-14%	2%
5,673	12,278	116%	3,962	7,851	98%	19.7%	9.6%	40	41	3%	2%
33,251	47,650	43%	17,214	23,113	34%	4.0%	4.5%	36	32	-11%	1%
26,299	33,959	29%	18,267	23,580	29%	8.5%	8.4%	33	29	-12%	1%
3,591	7,288	103%	2,182	4,134	90%	28.0%	24.9%	23	27	19%	1%
9,538	16,883	77%	3,572	6,515	82%	5.2%	5.4%	15	26	76%	1%
35,199	49,031	39%	17,607	24,451	39%	6.8%	8.4%	26	24	-8%	1%
30,056	43,095	43%	18,353	26,620	45%	8.2%	8.3%	22	23	4%	1%
553	1,359	146%	475	1,101	132%	10.1%	9.3%	19	19	-1%	1%
49,454	85,322	73%	24,889	38,471	55%	4.5%	3.5%	12	17	42%	1%
31,114	44,728	44%	17,830	26,655	49%	4.3%	4.4%	19	17	-14%	1%
44,442	67,285	51%	28,845	42,807	48%	4.1%	4.5%	15	14	-9%	1%
3,408	9,067	166%	2,191	5,108	133%	17.3%	7.8%	4	17	358%	1%
39,807	45,751	15%	19,916	24,666	24%	4.6%	13.7%	13	12	-6%	1%
1,048	3,012	187%	639	2,038	219%	9.1%	8.1%	5	10	87%	0.4%
17,686	27,000	53%	12,606	20,027	59%	9.8%	12.5%	12	10	-12%	0.4%
39,563	56,090	42%	17,899	25,770	44%	5.5%	7.4%	12	10	-12%	0.5%
4,637	11,834	155%	3,034	7,161	136%	7.4%	8.4%	6	11	104%	0.5%
31,533	44,622	42%	16,882	26,066	54%	9.0%	8.4%	16	10	-36%	0.4%
18,148	32,789	81%	10,327	16,503	60%	4.8%	6.5%	7	7	0%	0.3%
2,731	7,564	177%	1,612	4,535	181%	7.0%	7.3%	3	7	163%	0.3%
1,098	3,040	177%	767	1,799	135%	10.6%	7.3%	2	8	213%	0.3%
13,799	18,619	35%	7,918	12,333	56%	5.0%	5.2%	7	6	-13%	0.2%
4,568	10,119	122%	3,279	7,277	122%	10.5%	11.9%	3	4	62%	0.2%
17,773	28,991	63%	9,840	16,862	71%	10.7%	6.7%	5	4	-20%	0.2%
977	2,009	106%	688	1,491	117%	11.4%	7.4%	3	3	0%	0.1%
2,197	6,202	182%	1,604	4,230	164%	14.4%	11.8%	4	3	-13%	0.1%
1,671	2,848	70%	1,203	2,125	77%	11.9%	9.1%	1	6	743%	0.3%

Source: prepared by Markestrat data based on data from Euromonitor International and Tetra Pak, World Bank and CitrusBR,

Table 27b. Relationship among demographic data and consumption of orange-flavor drinks.

	Consumption of orange juice per capita (Liters FCOJ reconstituted as 100% juice at lowest allowable Brix of each market)			Juice (100% juice) Share of orange flavor (%)		Preferred flavors in 2010
	2003	2010	Variation	2003	2010	1st place
World						
Selected markets						
United States	18.44	13.99	-24%	61%	55%	Orange
Germany	17.52	13.67	-22%	36%	34%	Apple
France	14.30	15.73	10%	56%	55%	Orange
United Kingdom	13.29	12.43	-6%	65%	57%	Orange
Canada	19.34	18.98	-2%	43%	48%	Orange
Japan	4.12	3.38	-18%	32%	25%	Vegetable
Russia	1.99	3.33	67%	15%	18%	Apple
China	0.18	0.35	91%	55%	49%	Orange
Spain	5.80	5.89	2%	30%	30%	Orange
Brazil	1.32	1.24	-6%	76%	29%	Coconut
Mexico	1.56	1.53	-2%	52%	49%	Orange
Australia	12.04	11.55	-4%	66%	67%	Orange
South Korea	4.99	4.21	-16%	54%	55%	Orange
Poland	5.87	6.08	4%	34%	33%	Orange
Netherlands	12.65	11.04	-13%	45%	34%	Orange
Italy	3.28	2.73	-17%	28%	28%	Orange
South Africa	2.58	2.88	12%	71%	60%	Orange
Saudi Arabia	3.51	5.30	51%	34%	38%	Orange
Sweden	16.58	14.67	-12%	59%	50%	Orange
Belgium	12.12	12.00	-1%	54%	49%	Orange
India	0.10	0.08	-13%	40%	29%	Mango
Norway	14.53	19.29	33%	61%	65%	Orange
Austria	13.53	11.17	-17%	39%	42%	Orange
Switzerland	11.51	9.82	-15%	40%	40%	Orange
Argentina	0.53	2.25	328%	41%	86%	Orange
Ireland	18.19	15.14	-17%	69%	67%	Orange
Ukraine	0.65	1.26	95%	13%	11%	Others
Greece	5.96	5.11	-14%	33%	31%	Multifruit
Denmark	12.36	10.59	-14%	47%	42%	Orange
Chile	1.88	3.56	89%	89%	84%	Orange
Finland	17.37	10.80	-38%	66%	58%	Orange
New Zealand	10.44	9.77	-6%	45%	38%	Orange
Romania	0.71	1.89	168%	38%	50%	Orange
Indonesia	0.06	0.18	187%	36%	35%	Orange
Taiwan	1.55	1.32	-15%	18%	29%	Orange
Turkey	0.23	0.34	48%	35%	13%	Apple
Israel	4.37	3.13	-29%	75%	76%	Orange
Philippines	0.22	0.19	-13%	11%	11%	Pineapple
Colombia	0.45	0.35	-22%	92%	97%	Orange
Morroco	0.12	0.96	676%	69%	67%	Orange

Consumption of 66° Brix equivalent FCOJ does not include orange juice used to produce carbonated soft drinks: estimated 70,000 tons of FCOJ annually.

Nectar (25 to 99% juice)			Still drink (less than 25% juice)		
Share of orange flavor (%)		Preferred flavors in 2010	Share of orange flavor (%)		Preferred flavors in 2010
2003	2010	1st place	2003	2010	1st place
-	4%	Multifruit	26%	23%	Orange
22%	21%	Multifruit	24%	21%	Apple
43%	36%	Others	43%	33%	Orange
13%	13%	Cranberry	70%	34%	Orange
33%	15%	Others	24%	37%	Orange
37%	41%	Orange	40%	27%	Others
16%	23%	Mixed	15%	18%	Others
55%	55%	Orange	55%	38%	Orange
21%	20%	Others	67%	62%	Orange
16%	10%	Others	93%	90%	Orange
-	-	Mango	14%	16%	Orange
31%	30%	Orange	-	-	Blackcurrant
9%	23%	Tangerine	16%	26%	Pomegranate
13%	16%	Carrot	18%	14%	Multifruit
18%	7%	Tropical	12%	6%	Multifruit
8%	7%	Pear	50%	52%	Orange
62%	62%	Orange	69%	67%	Orange
30%	33%	Orange	33%	34%	Orange
55%	50%	Orange	44%	34%	Multifruit
46%	40%	Orange	71%	67%	Orange
-	-	Apple	1%	6%	Mango
40%	32%	Orange	65%	67%	Orange
38%	30%	Orange	16%	17%	Orange
40%	40%	Orange	41%	40%	Orange
26%	19%	Apple	53%	46%	Orange
4%	6%	Cranberry	64%	6%	Orange
13%	11%	Others	2%	2%	Multifruit
26%	21%	Multifruit	-	-	Peach
48%	40%	Orange	48%	43%	Orange
34%	31%	Orange	34%	49%	Orange
51%	48%	Orange	62%	57%	Orange
30%	19%	Tropical	16%	17%	Others
38%	27%	Orange	40%	34%	Orange
31%	29%	Orange	23%	35%	Orange
4%	6%	Mixed fruits & veg.	17%	22%	Others
8%	5%	Peach	18%	19%	Apricot
40%	39%	Orange	40%	27%	Orange
37%	35%	Orange	70%	64%	Orange
14%	7%	Mango	12%	17%	Blackberry
64%	64%	Orange	-	-	-

Source: prepared by Markestrat data based on data from Euromonitor International and Tetra Pak, World Bank and CitrusBR.

33. Orange flavor

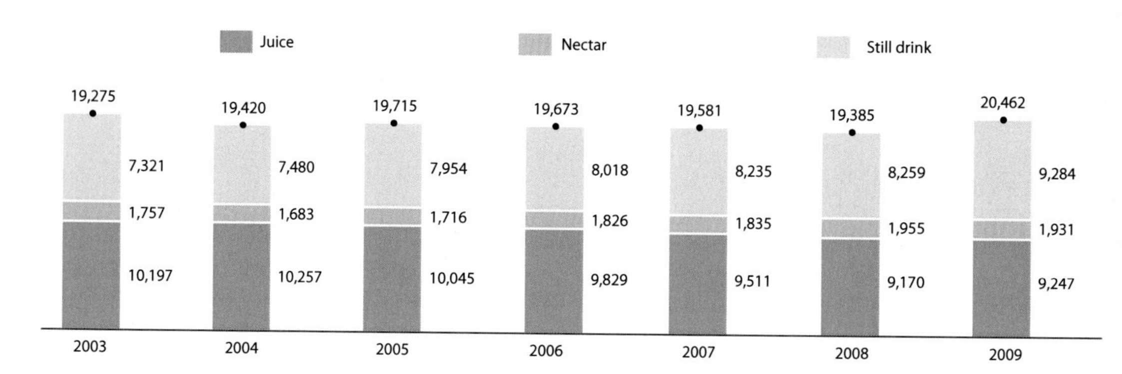

Graph 39. **Evolution in the consumption of orange juice in the selected countries, in millions of liters, per category of beverage.**

Source: prepared by Markestrat based on data from Tetra Pak and Euromonitor International.

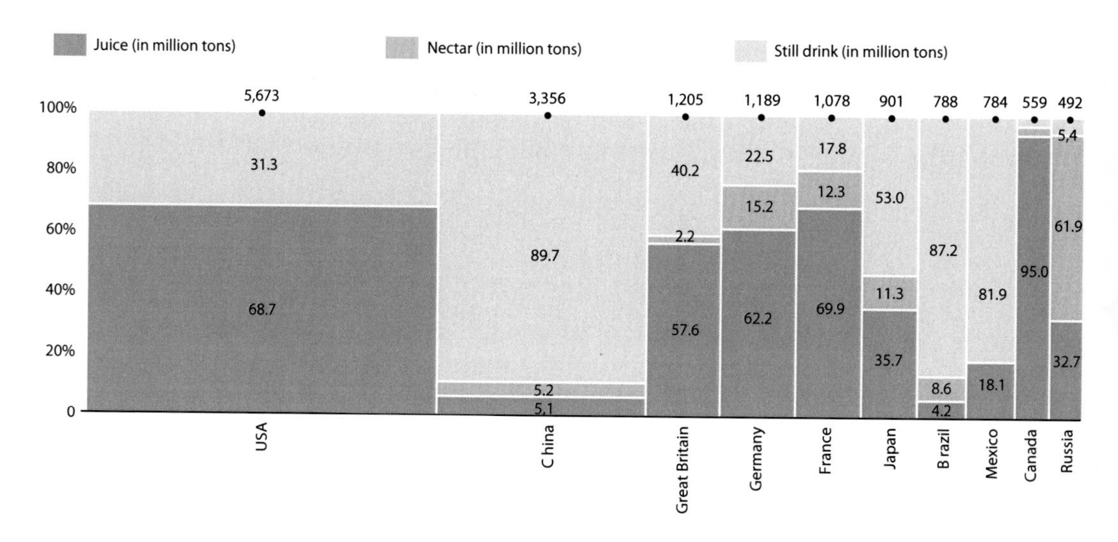

Graph 40. **Consumption of orange flavor per category of beverage in selected countries.**

The wider the bar, the greater the consumption of that country in relation to the other countries.

Source: prepared by Markestrat based on data from Tetra Pak and Euromonitor International.

34. Orange flavor in Europe

Europe, which is the main destination for Brazilian orange juice, consumed 29% of the global volume of orange flavored beverages in 2009, with 56% in the form of juices, 18% in nectars and 26% in still drinks (Graph 41). Between 2003 and 2009, their was a retraction of 2% in the consumption of orange drinks, for which juice was responsible, with a reduction of 7%, while nectar and still drinks presented increases in volume of 8% and 5% respectively. Of the 20 countries selected in Europe, 15 have orange as the preferred flavor in the category for juices, while 4 prefer apple.

Germany, with 1.2% of the world's population and a per capita net income of US$ 27.3 thousand, in 2009, presented a demand for 191 thousand tons of FCOJ Equivalent at 66° Brix. This is the largest destination market for Brazilian orange juice, since the United States is basically supplied by its own production. However, since the preferred flavor in Germany is apple, the relationship between the retail price of orange juice and apple juice has a major impact on the consumption of these products. Orange juice is always more expensive than apple juice (Graph 42), however, when the difference in price between the two increases, there is a migration of consumption from apple to orange flavor, especially among consumers with larger incomes.

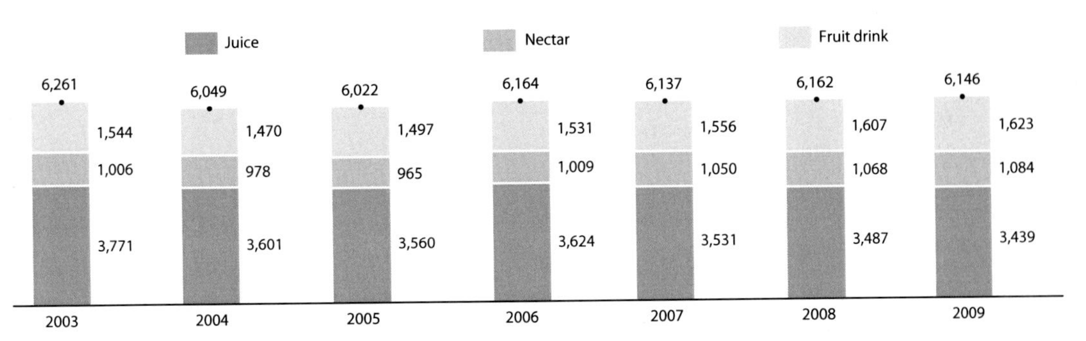

Graph 41. Evolution in the consumption of orange juice in Europe, in millions of liters, per category of beverage.

Source: prepared by Markestrat based on data from Tetra Pak and Euromonitor International.

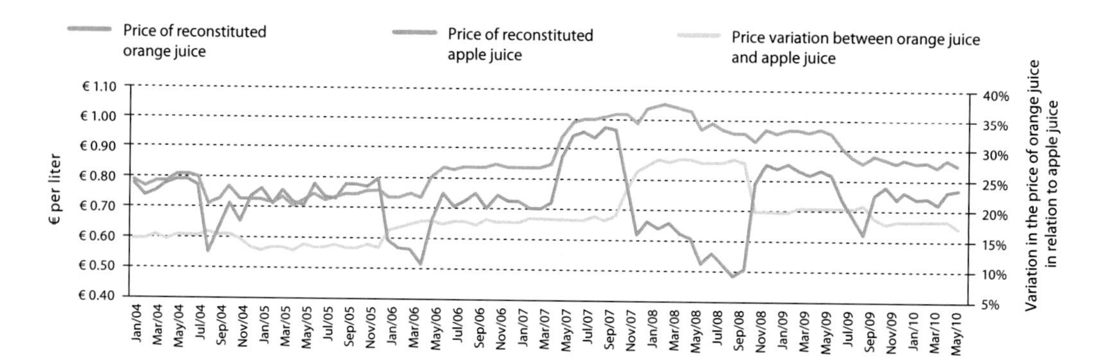

Graph 42. **Variation in the German retail price for orange juice in relation to apple juice.**
Source: prepared by Markestrat based on CitrusBR.

35. Orange flavor in North America

In 2009, North America demanded a volume of 6.2 billion liters of orange flavored beverages, 11.5% less than in 2003. In this market, consumption mostly occurs in the form of juice and, to a lesser degree, as still drinks. Nectar is an insignificant market. Of this total, the United States consumed 5.7 billion liters, which is equivalent to 92% of the total consumption for North America.

The United States, with 4.5% of the world's population and a per capita net income of US$ 32.9 thousand, represents the largest and most influential market for orange juice on the planet, since, besides being Brazil's primary competitor in the production of FCOJ, they are also the largest consumers of juice. With a demand in 2009 of 851 tons of FCOJ Equivalent at 66° Brix, the United States account for 38% of global consumption.

The demand for orange juice has dropped by 24% in the last decade, going from 1,114 thousand tons to 851 thousand tons, suffering a retraction of 263 thousand tons, equivalent to a drop in the annual demand of around 60 million boxes of oranges (Graph 43).

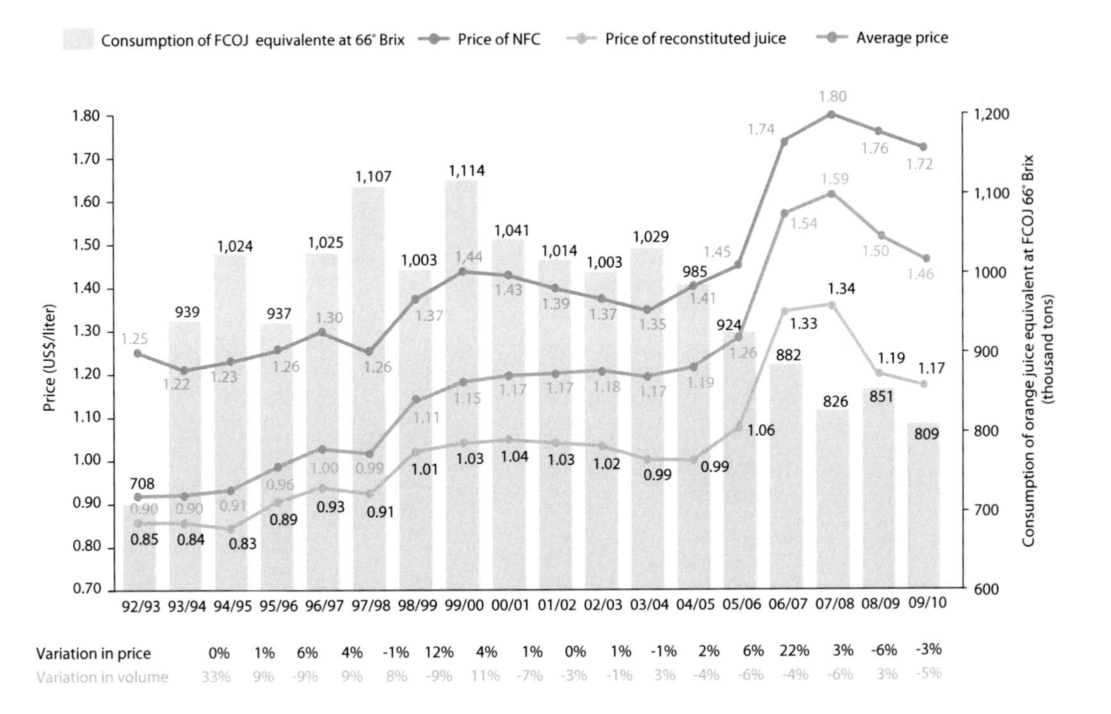

Graph 43. **Variation between price and consumption of orange juice in the United States.** The consumption equivalent of FCOJ at 66° Brix was calculated based on data for initial and final harvest stock (on the dates of September 30) and data for production, import and export (between October of year 1 and September of year 2), reported by the FDOC (Florida Department of Citrus). The prices were established based on the data from Nielsen, also reported by the FDOC. Source: prepared by Markestrat.

Such a reduction was the consequence of a number of factors. In 2000 consumer behavior began to undergo a change, due to the rise in the unemployment rate (previously standing at 4%), which was marked by the 09/11 terrorist attacks, leading to the first adjustment in the American economy for that decade.

In the middle of the first decade of 2000, the beverages sector began to feel the effect of the low carbohydrate diets that were being widely promoted in the United States – the Atkins and South Beach diets – with an increase in demand for products with fewer calories.

Later, in 2004 and 2005, the hurricanes arrived, reducing the supply and exponentially raising the price per box of oranges and, to a lesser degree, the prices for juice on the supermarket shelves. This disproportionality in the price paid for the fruit and in the selling price for the juice reduced the margins for the packaging companies on orange flavored beverages. This dynamic led to a cut in investments in the advertising and promotion of orange juice sales, thereby further intensifying the reduction in consumption.

Lastly, the recent global financial crisis, which raised the level of unemployment to 9.2% in the United States, meant that a percentage of the population stopped consuming more costly products, such as orange juice. This same period coincides with the acceleration of innovations in the beverages industry, coming to offer an avalanche of new products at lower costs, with higher profit margins, fewer calories and also having a more appealing and modern image.

As presented in Graph 44, in 2009 the United States consumed 5,673 million liters of orange beverages, including the categories of juices and still drinks, or 851 million tons of FCOJ equivalent at 66° Brix (Graph 44).

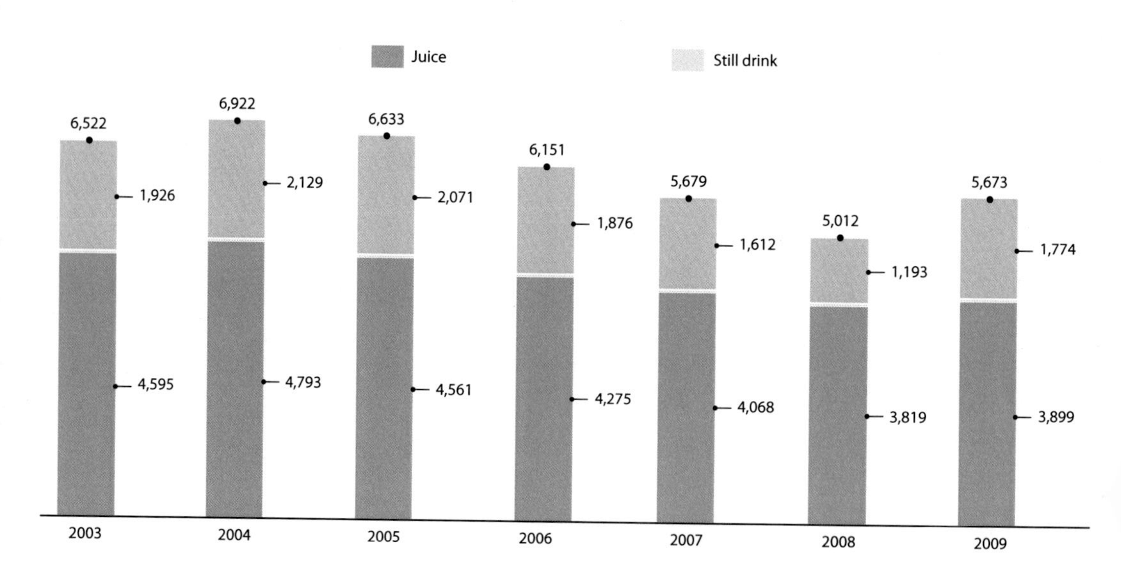

Graph 44. Consumption of orange flavor in the United States per type of beverage in millions of liters.

Source: prepared by Markestrat based on data from Tetra Pak and Euromonitor International.

In contrast to Europe, the United States produce a good part of the orange juice they consume. Of the 851 thousand tons of FCOJ Equivalent at 66° Brix consumed in the 2009/09 harvest, 88% was produced internally (Table 28).

With the economic advantages of the non-imposition of import duties and anti-dumping fees, Costa Rica and Mexico are leading countries in the supply of orange juice to the United States. Between the 1992/93 and 2008/09 harvests, these countries raised their exports of FCOJ equivalent at 66° Brix from 13 thousand tons to 86 thousand tons, thereby causing a drop, from 88% to 54%, of Brazilian share in American imports.

With the recovery of the Florida citrus groves following the hurricanes, the maintaining of imports and the recent reduction in consumption, American inter-harvest reserve stocks, in the last three years, rose from 16 to 29 weeks of consumption in FCOJ equivalent at 66° Brix.

Table 28. Balance of supply and demand of the United States.

		92/93	93/94	94/95	95/96	96/97	97/98
Initial stock on October 1	1000 tons	88	153	230	177	194	283
Internal production of orange juice							
Florida	1000 tons	596	741	849	850	974	1,044
Texas	1000 tons	0	0	1	1	2	2
California/Arizona	1000 tons	48	49	32	37	34	40
Total	1000 tons	644	789	882	887	1,010	1,086
Orange juice imports							
Belize	1000 tons	7	5	6	7	13	6
Brazil	1000 tons	187	256	104	104	137	124
Canada	1000 tons	1	0	1	1	1	1
Costa Rica	1000 tons	2	3	4	6	14	16
Dominican Republic	1000 tons	0	0	0	1	1	1
Honduras	1000 tons	3	2	4	4	6	3
Mexico	1000 tons	11	32	50	31	37	48
Other countries	1000 tons	2	2	2	1	0	0
Total	1000 tons	211	301	170	156	209	199
Share of Brazilian orange juice in orange juice imports to the United States		88%	85%	61%	67%	66%	62%
Orange juice exports							
Canada	1000 tons	29	18	22	24	28	31
Europe	1000 tons	30	29	39	39	54	48
Japan	1000 tons	8	14	4	11	10	13
Other countries	1000 tons	16	13	18	15	12	12
Total	1000 tons	83	74	82	89	105	103
Outstanding trade balance for orange juice	1000 tons	128	227	88	67	104	96
Final stock on September 30	1000 tons	153	230	177	194	283	358
Weeks of consumption		8	12	10	10	13	19
Attributed consumption in the United States	1000 tons	708	939	1,024	937	1,025	1,107

The equivalent consumption of FCOJ at 66 ° Brix was calculated based on data for initial and final harvest stock (on the dates of September 30) and data for production, import and export (between October of year 1 and September of year 2), reported by the FDOC (Florida Department of Citrus).

Source: prepared by Markestrat.

98/99	99/00	00/01	01/02	02/03	03/04	04/05	05/06	06/07	07/08	08/09	09/10
358	362	437	463	468	480	559	426	316	257	441	476
811	988	944	990	845	1,019	637	648	571	777	725	564
1	2	4	1	2	2	2	1	2	2	1	1
49	61	25	21	36	17	45	49	55	44	23	28
862	1,051	974	1,013	883	1,038	684	699	628	822	749	593
8	12	9	3	6	14	21	10	8	15	12	11
184	167	119	78	161	109	164	141	184	174	121	129
1	1	2	2	2	2	4	3	3	5	4	1
17	24	22	17	20	23	21	19	30	27	23	21
0	0	1	1	1	1	1	1	1	1	1	1
1	3	4	3	1	1	2	1	2	0	1	1
35	30	23	29	10	6	39	33	51	62	63	64
1	2	2	1	5	1	1	3	3	2	1	3
248	239	182	134	206	157	253	211	282	287	224	232
74%	70%	65%	58%	78%	70%	65%	67%	65%	61%	54%	56%
34	32	35	34	39	40	45	46	53	63	47	46
42	44	33	69	15	28	21	36	19	19	22	37
12	10	8	9	4	5	3	3	2	2	2	1
14	15	10	16	16	14	15	13	13	15	17	20
102	101	87	128	74	87	84	97	87	98	88	104
145	138	95	5	132	70	169	114	196	189	136	128
362	437	465	468	480	559	426	316	257	441	476	388
17	22	24	24	24	30	24	19	16	27	29	24
1,003	1,114	1,041	1,014	1,003	1,029	985	924	882	826	851	808

36. The orange flavor in the BRIC group countries plus Mexico

The region formed by the BRIC countries plus Mexico, with 43.4% of the world's population, consumes 221 thousand tons of FCOJ Equivalent at 66° Brix, with 80% in the form of still drinks, 10% in the form of nectar 10% as juice, and a consumption profile per category of beverage typical of countries with lower per capita net income. Between 2003 and 2009, the volume of orange flavored beverages (juice, nectar and still drinks) rose by 50%, with still drinks rising by 62%.

In 2009 China, with 19.6% of the world's population and a per capita net income of US$ 2.02 thousand, consumed 3.4 billion liters of orange flavored beverages (juice, nectar and still drink), a volume 64% higher than that in 2003, registering an average annual growth rate in juices of 13.3% and of 10% per year for still drink.

The nectar category, on the other hand, retracted by 6.3% per year. Despite the major growth rate in consumption registered in China, in 2009 they consumed just 74 thousand tons of FCOJ equivalent at 66° Brix, revealing that consumption is basically in the form of still drinks with low orange juice content (Graph 45).

In India, with 17.2% of the world's population and a per capita net income of US$ 823, the preferred flavor is mango. In that country, the consumption of FCOJ equivalent at 66° Brix in 2009 was just 19 thousand tons, or 236 million liters of orange flavored beverages, including still drink and juices (Graph 46).

Brazil, with 2.9% of the world's population and a per capita net income of US$ 5.23 thousand, consumed, in 2009, 41 thousand tons or 788 million liters of FCOJ Equivalent at 66° Brix in the form of industrialized juices, nectars and still drink (Graph 47).

With the exception of Brazil, the concentration of orange juice in nectars and still drink for the countries in the BRIC group + Mexico is much lower than that presented by the

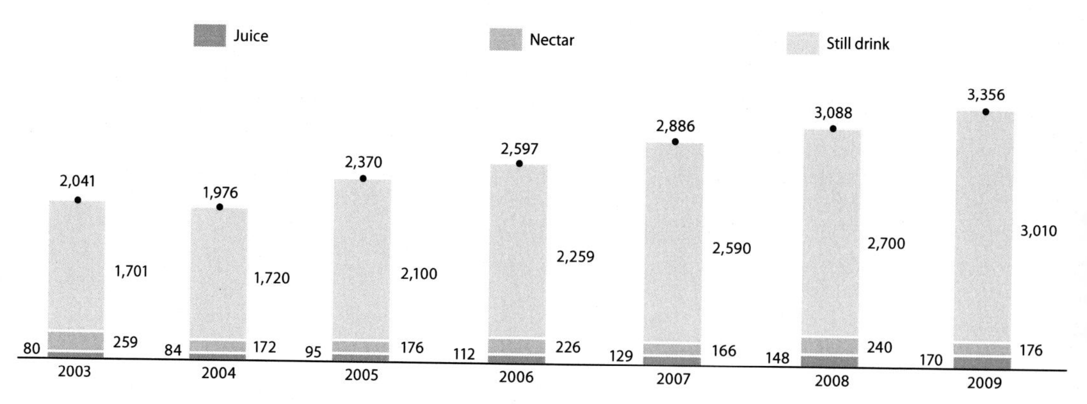

Graph 45. Evolution in the consumption of orange juice in China, in millions of liters, per category of beverage.

Source: prepared by Markestrat, based on TetraPak and Euromonitor International.

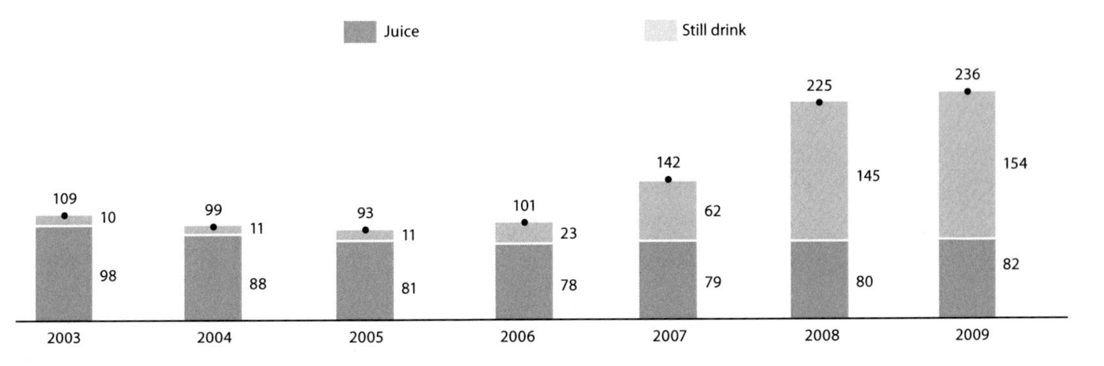

Graph 46. **Evolution in the consumption of orange juice in India, in millions of liters, per category of beverage.**

Source: prepared by Markestrat, based on TetraPak and Euromonitor International.

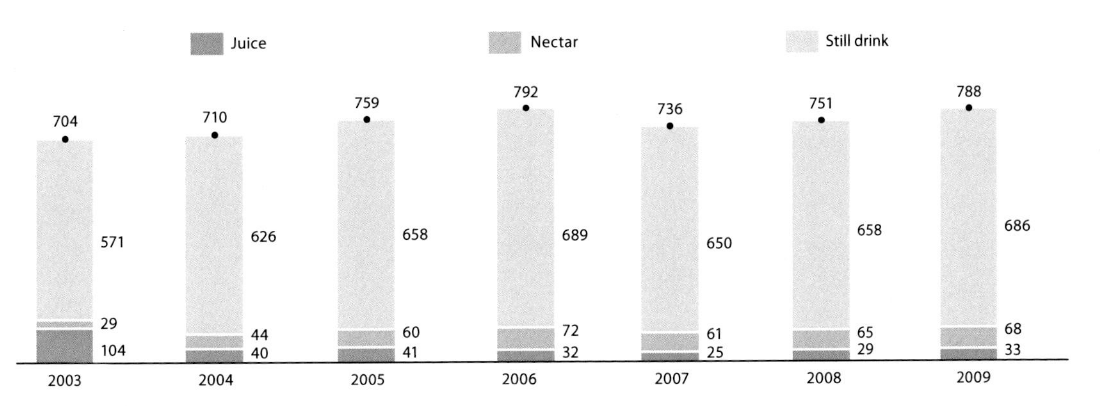

Graph 47. **Evolution in the consumption of orange juice in Brazil, in millions of liters, per category of beverage.**

Source: prepared by Markestrat, based on TetraPak and Euromonitor International.

countries in Europe. In Germany, orange flavored nectars and still drink contain, respectively, 72% and 11% juice content in the packaged product. In China these percentages drop to 25% and 5%, respectively.

Therefore, the market potential for the orange flavor in these emerging nations is related to the quality of the drink. If these countries had maintained the same volume of consumption for the orange flavor in 2009, of 5.6 billion liters, but migrated to the consumption of beverages diluted to the same proportions of juices, nectars and still drink consumed in Germany and the same dilution factors as that country, the increase in consumption would be around 142 million boxes of oranges (Table 29).

Table 29. Simulated analysis of consumption of juice in the bric group countries and Mexico.

Country	GDP per capita (US$)	Per capita net income (US$)	Juice content dilution factor		Segments	Consumption of orange juice in 2009	
			Nectar	Still drink	Juice / Nectar / Still drink	liters (×1000)	tons FCOJ equiv. to 66° Brix
Brazil	8,117	5,227	75%	20%		788	41,475
Russia	8,677	5,242	50%	13%		492	56,169
India	1,089	823		10%		236	18,246
China	3,697	2,025	25%	5%		3,356	68,248
Mexico	8,117	5,632		9%		784	37,412
BRIC + Mexico	5,991	3,790			500 / 549 / 4,519	5,656	221,552
Germany	40,824	27,299	72%	11%	62% / 15% / 23%		
BRIC + Mexico – simulated consumption					3,507 / 848 / 1,301	5,656	792,658
Additional demand for orange juice based on the simulation					Additional demand for 142 million boxes of oranges		571,105

Source: prepared by Markestrat, based on TetraPak and Euromonitor International.

37. The growth potential of Brazil's domestic market

In the 2009/10 harvest, the per capita consumption of orange juice in Brazil was 12.3 liters, when adding together the consumption of the 41 thousand tons of diluted FCOJ to the 100 million boxes of oranges sold for fresh consumption on the domestic market which are almost entirely transformed into freshly squeezed orange juice in bars, bakeries, restaurants, hotels and residential homes (Table 30).

If the consumption of orange juice in Brazil were on a par with the levels in countries that have a daily habit of consuming industrialized orange juice, the increase in demand for Brazilian oranges could be as much as 22 million to 65 million boxes. This demonstrates the need for government policies and strategies for private initiatives to better exploit the domestic market (Table 31).

Table 30. Current consumption of oranges and orange juice in Brazil.

	Total harvest 2009/10 Million boxes	Fresh orange consumption 2009/10 Million boxes	Consumption by industries 2009/10 Million boxes
São Paulo and Triângulo Mineiro (CitrusBr)	317.4	43.3	274.1
Bahia and Sergipe (IBGE)	44.0	35.4	8.6
Paraná and R.G. do Sul (IBGE)	13.1	4.6	8.5
Pará 2009-10 (IBGE)	5.0	5.0	0.0
Goiás 2009-10 (IBGE)	3.1	3.1	0.0
Rio de Janeiro (IBGE)	1.4	1.4	0.0
Other states (IBGE)	7.2	7.2	0.0
TOTAL BRAZIL	391.2	100.0	291.2

Consumption of fresh oranges:	in fruit	4,081,224,000 kg of fruit
	in juice equivalent	2,148,012,632 liters of juice
Consumption of industrialized juice in Brazil (41,000 tons of diluted FCOJ)		231,203,008 liters of juice
Total consumption of orange juice in Brazil (fresh orange + diluted FCOJ)		2,379,215,639 liters of juice
Brazilian population		192,876,397
Per capita consumption of orange juice in Brazil		12.3 liters of juice

Source: prepared by Markestrat, based on data from the IBGE and CitrusBr.

Table 31. Potential for growth in the domestic market putting it on a par with the Brazilian per capita consumption to those countries where the population has a habitual daily consumption of industrialized orange juice.

Selected countries	Liters of orange juice per capita	Potential for increase in Brazil	
		Liters	Boxes of oranges
Norway	20.0	1,477,737,980	65,852,851
United States	17.2	944,789,989	42,102,941
Ireland	17.1	916,204,515	40,829,078
Canada	16.7	838,087,101	37,347,910
France	14.9	500,834,989	22,318,850

Source: prepared by Markestrat.

38. The power of international retail

Brazilian industries have become specialized in the production and international distribution of orange juice in the same way that the packers and major brands have specialized in the bottling and sale of beverages to the retail sector in each of the countries where they operate.

In order to gain efficiency in this highly competitive market, all of the links in the productive chain for orange juice are becoming increasingly concentrated, from the supply producing companies to the retail distribution channels (Table 32 and Figure 8). Because of this, the bargaining power of retailers with the packaging companies and these, in turn, with the Brazilian orange juice exporters, is out of proportion (Table 33). The growth of the major chains, year after year, in the sale of foodstuffs is notorious. Walmart alone sold more than US$ 425 billion in 2009 (Table 34).

Table 32. Concentration in the sale of food for the 5 top retailers in selected countries.

Countries	Market share		
	2000	2005	2010
Israel	99.3%	99.5%	100.0%
Switzerland	80.7%	85.1%	92.1%
South Korea	58.5%	72.3%	84.4%
Austria	72.5%	71.9%	84.4%
Germany	66.4%	72.9%	80.0%
France	70.0%	64.8%	74.7%
Russia	60.9%	55.1%	74.4%
Canada	60.6%	54.8%	73.7%
Japan	66.6%	63.4%	66.5%
Spain	52.7%	56.7%	69.2%
United Kingdom	50.6%	59.8%	67.9%
Italy	69.6%	67.5%	67.1%
Poland	51.4%	41.6%	53.2%
United States	42.7%	45.3%	46.3%
Brazil[1]	41.0%	40.5%	43.0%

[1] Data for Brazil is from the ABRAS and refers to the share of 5 largest retail chains on the self-service market.

Source: prepared by Markestrat based on Tetra Pak, IGD and Planet Retail.

1. With the aim of obtaining greater revenues per square meter and better operational efficiency, retailers control the availability of space for each type of product on the supermarket, giving preference to products with higher stock turnover and larger profit margins.

2. In addition, the retail chains reduce the number of items available on the shelves, thereby diminishing stock volumes and the complexity of the shopping process. This strategy brings logistical advantages, since the retailer gains in agility and reduces overall operating costs. The upshot of this is that the industries have less shelf space on which to expose their brands, further increasing the costs of the so-called slotting fees, which are fixed fees paid by the manufacturers to ensure exposure of their products on the supermarket shelves.

1. Increased operational efficiency and simplified shopping process

2. Reduction in costs and expenditures

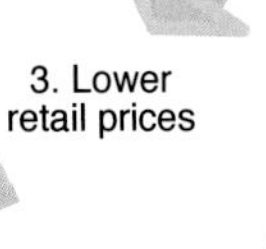
3. Lower retail prices

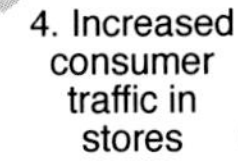
4. Increased consumer traffic in stores

5. Greater sales volumes

5. Lastly, the retail chains in Europe and the United States, the main destinations for Brazilian orange juice, operate with net margins that are significantly, in some cases up to three times, higher than those registered by the sector in Brazil.

3. Another strategy for reducing retail costs and expenditures is the adoption of own brands, the so-called private labels or white brands. This type of product is systematically gaining ground on the market in relation to the traditional brands, since, in the developed nations, they offer the same standard of quality at a lower price. Consequently, we see manufacturers of traditional brands putting enormous pressure on all the links in the chain in order to ensure the market competitiveness of their products.

4. Information technology is a fundamental tool used currently by retailers in the quest for efficiency and reduced purchasing costs. In the case of negotiations for juices in general, it has become a market practice for retail chains to invite the packaging companies, their suppliers of packaged juice, to participate in electronic reverse auction processes, where the supplier with the lowest price is able to close the deal with the distribution network. The use of information technology has also helped the end consumer to seek alternatives that make their shopping lists cheaper. In the United Kingdom, with the help of a single website (www.mysupermarket.co.uk), the consumer is able to compare food prices in their local main supermarket chains, before leaving home, thereby choosing which items to buy in each supermarket.

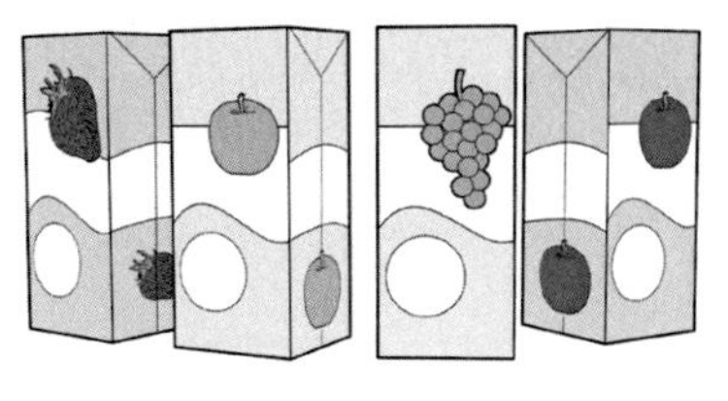

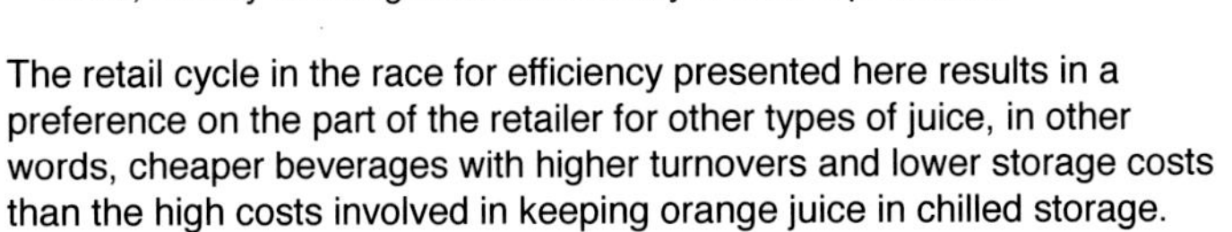

The retail cycle in the race for efficiency presented here results in a preference on the part of the retailer for other types of juice, in other words, cheaper beverages with higher turnovers and lower storage costs than the high costs involved in keeping orange juice in chilled storage.

Figure 8. The retail cycle in the race for efficiency and consequences for oranges.

Source: Markestrat, based on TetraPak.

Table 33. Share of the five main retailers that distribute orange juice in the main consumer markets.

Main consumer markets	Orange juice consumption per market			Share of the 5 main retailers per market				
				Share of the 5 largest retailers			Ranking according to total revenues	
							1st place	
	Tons FCOJ equivalents to 66° Brix	Share in world consumption	Accumulated share	In the total retail market	In the sale of foodstuffs	In the sale of non-alcoholic beverages	Sales billion US$	Retailer
Total of selected	2,267	100%						
1 United States	851	38%	38%	53%	47%	62%	320.8	Walmart
2 Germany	191	8%	46%	76%	77%	80%	59.1	Edeka
3 France	165	7%	53%	73%	68%	71%	63.5	Carrefour
4 United Kingdom	138	6%	59%	63%	60%	64%	64.7	Tesco
5 Canada	105	5%	64%	61%	59%	62%	23.7	Loblaw
6 Japan	75	3%	67%	65%	64%	-	72.1	AEON
7 Russia	74	3%	71%	68%	67%	-	9.9	X5 Retail Group
8 China	74	3%	74%	44%	47%	-	9.6	China Res, Ent,
9 Spain	47	2%	76%	72%	69%	68%	21.6	Mercadora
10 Brazil	41	2%	78%	41%	41%	-	12.8	Carrefour
11 Mexico	40	2%	79%	82%	83%	-	20.2	Walmart
12 Australia	40	2%	81%	92%	92%	-	38.4	Woolworths
13 South Korea	38	2%	83%	83%	77%	-	10.8	Shinsegae
14 Poland	37	2%	85%	50%	51%	53%	5.5	Jerónimo Martins
15 The Netherlands	33	1%	86%	59%	60%	63%	14.8	Ahold
16 Italy	29	1%	87%	66%	66%	68%	18.4	Coop Italia
17 South Africa	27	1%	88%	90%	90%	-	7.5	Shoprite
18 Saudi Arabia	23	1%	89%	78%	77%	-	1.9	Panda
19 Sweden	23	1%	90%	90%	90%	-	13.5	Ahold
20 Belgium	23	1%	91%	78%	80%	77%	8.7	Colruyt
21 India	19	1%	92%	70%	66%	-	1.4	Pantaloon
22 Norway	17	1%	93%	94%	93%	-	10.5	NorgesGruppen
23 Austria	17	1%	94%	79%	80%	79%	9.5	Rewe Group
24 Switzerland	14	1%	94%	86%	89%	89%	16.5	Migros
25 Argentina	13	1%	95%	68%	88%	-	6.3	Musgrave
26 Ireland	13	1%	96%	78%	76%	73%	3.8	Carrefour
27 Ukraine	12	1%	96%	67%	78%	-	1.4	Fozzy
28 Greece	11	0%	97%	66%	76%	77%	4.1	Carrefour
29 Denmark	11	0%	97%	86%	91%	-	9.7	Coop Danmark
30 Chile	11	0%	98%	92%	98%	-	4.3	Cencosud
31 Finland	9	0%	98%	89%	93%	-	13.5	SOK
32 New Zealand	7	0%	98%	94%	95%	-	6.3	Foodstuffs
33 Romania	7	0%	99%	75%	78%	76%	3.1	Metro Group
34 Indonesia	7	0%	99%	67%	84%	-	1.3	Indomaret
35 Taiwan	6	0%	99%	61%	64%	-	4.0	Pres, Chain Store
36 Turkey	6	0%	99%	68%	74%	-	4.2	Migros Ticaret
37 Israel	4	0%	100%	100%	100%	-	3.2	Shufersal
38 Morroco	4	0%	100%	94%	93%	-	1.2	ONA
39 Philippines	3	0%	100%	82%	80%	-	2.1	SM Investment
40 Colombia	3	0%	100%	84%	90%	-	2.9	Casino

Consumption of FCOJ equivalent at 66° Brix not including orange juice used for the production of fizzy drinks: an estimated 70,000 tons per year of FCOJ.

2nd place		3rd place		4th place		5th place	
Sales billion US$	Retailer	Sales billion US$	Retailer	Sales billion US$	Retailer	Sales billion US$	Retailer
82.5	Kroger	66.7	Target	63.8	Walgreens	57.6	Costco
41.1	Rewe Group	38.1	Schwarz Group	33.6	Aldi	26.7	Metro Group
44.4	Leclerc	37.6	Casino	35.5	ITM (Intermarché)	31.5	Auchan
32.8	Sainsbury	31.9	Walmart	25.5	Morrisons	15.1	Marks &Spencer
16.9	Walmart	14.4	Sobeys	10.3	Costco	10.2	Metro (CAN)
58.9	Seven &I	20.7	Uny	18.5	LAWSON	17.4	Isetan Mitsukoshi
6.1	Magnit	5.6	Auchan	4.9	Metro Group	2.4	O'Key
9.3	Lianhua	7.4	Auchan	6.0	Walmart	5.6	Carrefour
20.3	Carrefour	20.0	El Corte Inglés	11.8	Eroski	6.8	Auchan
11.0	Casino	10.9	Walmart	2.7	Lojas Americanas	2.6	SHV Makro
7.1	Soriana	4.0	OXXO	3.2	Comercial Mexicana	3.1	Chedraui
28.7	Coles Group	14.4	Metcash (AUS)	2.4	Aldi	1.9	AUR
9.8	Lotte Shopping	7.0	Tesco	3.9	GS Retail	2.0	Eland
4.9	Metro Group	3.3	Tesco	3.1	Schwarz Group	2.8	Carrefour
5.2	C1000	3.8	Aldi	3.6	Sperwer	3.4	Sligro
13.0	Auchan	10.9	Carrefour	9.9	Conad	8.4	Esselunga
6.0	Pick n Pay	3.7	SPAR (South Africa)	3.5	Massrnart	2.3	Metcash (RSA)
1.0	Bin Dawood	0.9	Carrefour	0.8	Al Othaim	0.4	Al Sadhan
7.4	KF Gruppen	6.8	Azei Johnson	3.5	Systembolaget	3.2	Apoteket
7.9	Carrefour	7.1	Delhaize Group	3.9	Aldi	2.8	Louis Delhaize
0.6	Reliance Retail	0.3	Aditya Birla	0.3	RPG Group	0.2	Metro Group
6.1	Reitan	4.2	Coop Nome	3.7	Ahold	1.4	Phoenix
6.9	SPAR (Austria)	4.6	Aldi	1.3	Schwarz Group	1.2	Metro Group
13.6	Coop (CH)	2.7	Manor	1.5	Rewe Group	1.1	Aldi
4.1	Tesco	3.2	Dunnes	2.3	Stonehouse	1.1	La Anónima
2.5	Cencosud	1.9	Walmart	1.5	Coto	2.1	BWG
1.3	Metro Group	1.1	ATB Market	0.8	Furshet	0.6	Retail Group
2.3	Schwarz Group	2.2	Delhaize Group	1.8	Sklavenitis	1.4	Veropoulos
7.8	Dansk Supermarked	5.5	Dagrofa	2.5	SuperBest amba	1.6	Reitan
3.6	Walmart	2.2	SMU	0.7	Falabella	0.7	FASA
7.8	Kesko	2.0	Suomen Lähikauppa	1.1	Wihuri	1.1	Stockmann
3.9	Woolworths (AUS)	1.1	The Warehouse Group	0.2	LVMH	0.2	Coles Group
2.0	Rewe Group	1.7	Carrefour	1.3	Schwarz Group	0.6	Louis Delhaize
1.2	Carrefour	0.7	Matahari	0.7	Dairy Farm	0.7	Alfa Mart
2.7	Isetan Mitsukoshi	1.9	Carrefour	1.5	Auchan	1.4	PXmart
3.8	BIM	2.2	Carrefour	1.8	Metro Group	1.0	Tasco
2.0	Blue Square	0.7	Tiv Taam	0.0	Delek	0.0	L'Occitane
0.3	Metro Group	0.3	Casino	0.2	Hanouty	0.1	Label'Vie
0.7	Puregold	0.5	Robinsons	0.5	Mercury Drug	0.4	China Res. Ent.
1.9	Carrefour	0.8	Olimpica	0.5	LA 14	0.4	Alkosto

Source: prepared by Markestrat based on TetraPak, IGD and Planet Retail.

Table 34. World retail ranking in 2010.

Retail position	Retailer	Region	Total revenues (billion dollars)	Sales of foodstuffs (billion dollars)	Number of outlets
1	Walmart	Worldwide	446,506	200,459	8,960
2	Carrefour	Worldwide	148,771	102,678	15,978
3	Tesco	Europe, Asia	104,351	67,578	5,381
4	Metro Group	Europe, Russia	102,100	40,859	2,215
5	AEON	Asia	95,734	57,161	14,485
6	Seven & I	Asia, Oceania, North America	93,089	51,690	25,031
7	Kroger	United States	86,151	64,061	3,614
8	Schwarz Group – Lidl Kaufland	Europe, Asia	85,262	67,353	10,439
9	Costco	North & Central Americas, Asia	79,260	41,915	573
10	Auchan	Europe, Asia, Africa	78,987	48,145	3,049
11	Rewe Group	Europe	74,572	50,168	13,331
12	Casino – Extra, Pão de Açucar	Europe, South America, Africa	74,507	44,718	11,253
13	Aldi	Europe, United States	72,615	61,850	9,617
14	Target	United States	68,562	14,116	1,760
15	Sears	North & Central Americas	66,224	4,882	3,845
16	Ahold	Europe, United States	63,116	50,242	5,225
17	Edeka	Germany	58,397	50,314	15,198
18	Woolworths (AUS)	Oceania, India	55,928	41,985	3,943
19	Leclerc	Europe	49,692	30,442	1,157
20	Coles Group	Australia, New Zealand	47,878	28,275	3,408
21	Safeway	North America	44,310	34,452	1,886
22	ITM (Intermarché)	Europe	42,574	33,025	3,539
23	Sainsbury	United Kingdom	35,430	23,091	934
24	SuperValu	United States	34,748	28,203	2,450
25	Loblaw	Canada	32,927	18,441	1,442
26	Delhaize Group	Europe, United States	29,633	24,173	2,798
27	Système U	Europe, South America, Africa	27,090	21,485	1,441
28	Morrisons	United Kingdom	26,975	23,816	439
29	Migros	Europe	26,702	12,626	2,052
30	Publix	United States	26,174	20,626	1,088

Source: Planet Retail - 15 08 2011.

Even the smallest retailers have organized themselves into purchasing pools or organizations in order to raise their bargaining power to be able to compete with the largest retail chains. The AMS, the largest purchasing pool, consists of 12 retail chains and purchases more than Carrefour throughout the world, or five times the aggregate revenues of the top 5 retail chains in Brazil. Coopernic, EMD and Agenor/Aludis individually purchased more than Tesco, the largest chain in the United Kingdom. The revenues for these pools is higher than the revenues for the largest retail chains in Brazil (Table 35).

Table 35. Purchasing pools or organizations set up by retailers in Europe and revenues for the five main retail chains in Brazil.

Purchasing pools or organizations set up by retailers in Europe		Estimated revenues for the members in 2010 (million US$)
Pool/organization	**Members**	
AMS 19 countries	Ahold Booker Dansk Supermarked Delhaize Esselunga ICA Jerónimo Martins Kesko Migros Morrisons Superquinn Système U	235,205
COOPERNIC 23 countries	Rewe Group Conad E.Leclerc Coop Schweiz Colruyt	165,365
ALADIS 10 countries	Edeka Eroski Intermarché	112,121
EMD 19 countries	Axfood Euromadi Markant SuperGros Musgrave Group Norgesgruppen Tuko Logistics Superunie ESD Italie Mercator	109,885
BIGS 14 countries	SPAR franchise holders in: Austria, Belgium, Czech Republic, Denmark, Finland, Greece, Hungary, Ireland, Italy, Slovenia, the Netherlands, Switzerland, UK	36,074

Table 35. Continued.

Purchasing pools or organizations set up by retailers in Europe		Estimated revenues for the members in 2010 (million US$)
Organization	**Members**	
BLOC 4 countries	Cactus Cora Louis Delhaize Delberghe Deli XL Distri-Group 21 Frost Invest Hanos Nederland HMIJ EUG Huyghebaert HorecaTotaal Lambrechts La Provencale LDIP Maximo Theunissen VAC	17,176
CBA 11 countries	Independent retailers and wholesale grocery businesses in: Bosnia-Herzegovina, Bulgaria, Croatia, Hungary, Latvia, Lithuania, Montenegro, Poland, Romania, Serbia, Slovakia, Slovenia	
CRAI 19 countries	Independent retailers and wholesale grocery businesses in: Albanian Republic, Italy, Malta, Switzerland and other countries	
	Retailer	**Total revenues 2009 (million US$)**
Five main retailers in Brazil	Casino – Extra, Pão de Açúcar	20,450
	Carrefour	16,414
	Walmart	12,636
	Lojas Americanas	6,106
	SHV Makro	3,321

Website for consumers in the United Kingdom to check prices of the items on their lists at all the supermarkets in their neighborhoods before going out to do their shopping www.mysupermarket.co.uk

Source: prepared by Markestrat based on TetraPak, IGD and Planet Retail.

39. Concentration of juice bottlers

Because of the redesign of retail and the consolidation of leadership for the major chains, a number of bottling companies have merged in order to increase scale and survive the growing economic pressures. In the United States, for example, the four top packaging companies hold 75% of the market, in the United Kingdom this rate rises to 84% (Table 36). In the last decade, it is estimated that more than 100 bottling companies have gone under and sold their businesses, diminishing the quantity of potential customers for the Brazilian orange juice industry to commercialize its production by 20%.

As part of the strategy for diversification, the bottling companies package and distribute a variety of other beverages, including other fruit juices, fizzy drinks and isotonic, teas, milk drinks, and waters, thereby making full use of the existing manufacturing structure. These packaging companies give preference to the bottling/cartooning of beverages that generate higher turnover and greater profit margins, either because of the costs of the raw material, or the lower juice content, or even because of an alternative product that makes it possible to sell it at a higher price. In times of tight margins, this criterion becomes even more important.

FCOJ at the price of US$ 700 per ton proves more competitive than all the other flavors. While at US$ 1,500 per ton, FCOJ becomes more expensive than the apple flavor. At US$ 2,000 per ton, it is also less competitive than the pear, tangerine and white grape flavors. And at US$ 2,500 per ton, besides these, FCOJ is less competitive than the peach, grapefruit, lime, black grape and strawberry flavors. This can be seen in Table 37, which demonstrates the competitiveness of orange juice at different prices in relation to the other fruit juices at market prices for May 2010 for the raw material delivered to warehouses in Rotterdam, after the payment of import duties.

Table 36. Concentration of the orange juice distributors in the main consumer markets in 2009.

Main consumer markets for orange juice		Consumption per market		
		Consumption in tons of FCOJ equivalents at 66° Brix	Share in world consumption of orange juice	Accumulated share
Total selected countries		2,287	100%	
1	United States	851	38%	38%
2	Germany	191	8%	46%
3	France	165	7%	53%
4	United Kingdom	138	6%	59%
5	Canada	105	5%	64%
6	Japan	75	3%	67%
7	Russia	74	3%	71%
8	China	74	3%	74%
9	Spain	47	2%	76%
10	Brazil	41	2%	79%
11	Mexico	40	2%	79%
12	Australia	40	2%	81%
13	South Korea	38	2%	83%
14	Poland	37	2%	85%
15	The Netherlands	33	1%	86%
16	Italy	29	1%	87%
17	South Africa	27	1%	88%
18	Saudi Arabia	23	1%	89%
19	Sweden	23	1%	90%
20	Belgium	23	1%	91%
21	India	19	1%	92%
22	Norway	17	1%	93%
23	Austria	17	1%	94%
24	Switzerland	14	1%	94%
25	Argentina	13	1%	95%
26	Ireland	13	1%	96%
27	Ukraine	12	1%	96%
28	Greece	11	0%	97%
29	Denmark	11	0%	97%
30	Chile	11	0%	98%
31	Finland	9	0%	98%
32	New Zealand	7	0%	98%
33	Romania	7	0%	99%
34	Indonesia	7	0%	99%
35	Taiwan	6	0%	99%
36	Turkey	6	0%	99%
37	Israel	4	0%	100%
38	Morroco	4	0%	100%
39	Philippines	3	0%	100%
40	Colombia	3	0%	100%

Consumption of FCOJ Equivalent at 66° Brix, not including orange juice used for the production of fizzy drinks, for which the estimated quantity is 70,000 tons per year of FCOJ.

Accumulated share of the four main orange juice packaging companies per market				
Share of the four main packaging companies	Ranking per volume purchased			
	1st place bottler	2nd place bottler	3rd place bottler	4th place bottler
75%	Pepsico	Coca-Cola	Florida's Natural	Dean foods
61%	Stute	Eckes	Gerber-Emig	Wesergold
52%	PepsiCo	Leiterie Saint Dennis	Eckes	Refresco
84%	Gerber-Emig	Princes	PepsiCo	Britvic
81%	Coca-Cola	PepsiCo	Lassonde	Joriki
52%	Kirin	Ehimi	Nippon Milk	Coca-Cola
96%	PepsiCo	Coca-Cola	Winn-bill-Dann	Nidan
82%	Huiyuan	Uni-President	Coca-Cola	Ting Hsin
82%	Garcia Carrion	Still drink	Pascoal	Antonio Munoz
77%	Coca-Cola	Ambev	Schin	Global
55%	Coca-Cola	Jumex	Tampico	-
83%	National foods	P&N	Heinz	Grove
91%	Lotte	Woongjin	Coca-Cola	Maeil Dairy
87%	Sokpol	Maspex	Hortex	Agrosnova
98%	Still drink	Friesland	Passina	Vitality
45%	Conserva Italia	San benedetto	Zuegg	Parmalat
64%	ShopRite	Pick'n Pay	-	-
38%	Almarai	Al othman	Kuwait Danish Dairy	Alrawabi
90%	Skane	Hellefords	Kivics	Procordia
70%	Still drink	Sunnyland	Konings	inex
70%	Dabur	PepsiCo	Coca-Cola	-
92%	Tine	Nen	Danica	Lerum
92%	Rauch	Pfanner	Spitz	Pago
91%	Migros	Coop	Ramseier	Mittelland
30%	Baggio	Coca-Cola	Litoral Citru	-
95%	Mulrines	Batchelors	Obrian	-
79%	PepsiCo	Vitmark	Coca-Cola	Prodential
99%	Hellenic	Vivartia-Delta	Sparti Hellas	Aspi
82%	Arla-Rynkeby	Coro	Orana	-
38%	Emboteladoras	Soprole	Vital	Watts
90%	Valio	Still drink	Eckes	-
92%	Frucor	Coca-Cola	Simply Squeeze	Natural Dairy
69%	Tymbark-Maspex	Coca-Cola	Euro Drinks	QAb
100%	Coca-Cola	Ultra jaya	Diamond foods	Others
83%	Ting Hsing	Uni-President	Hey Song	Agv
46%	Cappy	Dimes	Tamek	Aroma
100%	Gan Samuel	Gat foods	-	-
-	-	-	-	-
40%	Coca-Cola	Del Monte	Dole	
40%	Gaseosas	Coca-Cola	Meals	Jugos Sas

Source: prepared by Markestrat based on TetraPak, IGD and Planet Retail!

Table 37. Competition between orange juice and other fruit flavors based on the market prices for May 2009.

Brix for the concentrated product	Brix for bottled product	Range of market price US$/ton		Import duty rate	Additional amount in import duty	Flavor	Price
66.0	11.2		2,500	12.2%		orange	€ 0.36
66.0	11.2		2,000	12.2%	256	orange	€ 0.28
66.0	11.2		1,500	12.2%	171	orange	€ 0.21
66.0	11.2		700	12.2%	85	orange	€ 0.10
	70.0			11.2%	0	sour apple	€ 0.15
70.0	11.2	1,050	1,000	25.5%	255	sweet apple	€ 0.16
70.0	11.9	1,250	1,100	0.0%	0	pear	€ 0.22
				11.2	0	tangerine	€ 0.25
65.0	15.9	1,025	975	0.0%	0	EU white grapes	€ 0.27
32.0	10.0	1,200	1,100	19.2%	211	EU pure peach	€ 0.29
58.0	10.0	2,100	1,900	0.0%	0	grapefruit HR	€ 0.30
					0	lime 500 GPL	€ 0.32
58.0	15.9	2,100	2,000	12.0%	240	black grape	€ 0.34
65.0	7.0	3,500	3,200	0.0%	0	strawberry	€ 0.34
60.0	12.8	2,000	1,900	0.0%	0	frozen pineapple	€ 0.37
32.0	11.2	1,000	900	19.2%	173	puretrans apricot	€ 0.37
	65.0	1,407		14.0%	0	chokeberry	€ 0.41
65.0	13.5	2,500	1,950	0.0%	0	sourcherry	€ 0.43
66.0	10.0	2,500	2,200	0.0%	0	redcurrant	€ 0.43
65.0	8.8	3,700	3,300	0.0%	0	black mulberry	€ 0.49
22.0	20.0	620	600	0.0%	0	pure banana	€0.49
				8.5%	0	pink guava	€ 0.50
65.0	10.0	4,200	3,900	0.0%	0	elderberry	€ 0.66
	8.0	4,800	4,500	10.9%	491	lemon 500 GPL	€ 0.66
28.0	15.0	1,400	1,250	3.8%	48	totapuri mango	€ 0.67
65.0	11.6	4,200	3,900	0.0%	0	blackcurrant	€ 0.71
66.0	12.0	4,200	3,900	0.0%	0	pomegranate	€ 0.76
65.0	7.0	10,500	9,800	0.0%	0	raspberry	€ 1.16
17.0	15.0	1,550	1,300	3.8%	49	alphonso mango	€ 1.18
65.0	10.0	6,500	6,000	0.0%	0	cranberry	€ 1.21
65.0	10.0	12,000	11,000	0.0%	0	blueberry	€ 1.92
52.0	13.5	8,500	8,000	0.0%	0	cloudy passion fruit	€ 1.96

Source: prepared by Markestrat based on CitrusBR and Dhálar.

40. Concentration in the Brazilian orange juice industry

As with other sectors of the world economy, the citrus farming industry has being consolidating itself over the course of time. This type of concentration is also witnessed in other sectors of Brazilian agribusiness, such as beef and pork products, pulp and paper, sugarcane and chicken, among others. This tendency is also present in the banking, automobile, mining and retail sectors.

The consolidation of the processors is justified by the quest for gains in efficiency generated by the economy of scale, such as, for example, the dilution of fixed costs, possibilities for setting up an efficient system for bulk storage and maritime shipping, as well as access to capital at competitive rates. However, the concentration of the processors does not happen in isolation; there are the links before and after the juice industry. The concentration of retailers is excessively significant. In Germany, for example, the five top retailers control 80% of the sales of non-alcoholic beverages. In turn, the juice bottlers, who are direct customers for the orange juice exported by Brazil, follow in the same direction. Today, just 35 bottlers buy up 80% of the world's production of orange juice, with the remaining 20% being bought by around 565 bottlers.

Also within the same trend, and seeking gains in efficiency as a result of greater scale, the producers have been swiftly consolidating, 2% of them already own 55% of the trees in the citrus belt.

ANTONIOSI

Closing message

Before the closing words of this study, I would like to invite the citrus farming community, language barriers aside, to travel to Argentina and get in touch with the lyrics of this composition written by Leon Gieco, which is known to us in the fondly remembered voice of Mercedes Sosa.

Sólo le pido a Dios
que el dolor no me sea indiferente,
que la reseca muerte no me encuentre
vacio y solo sin haber hecho lo suficiente.
Sólo le pido a Dios
que lo injusto no me sea indiferente,
que no me abofeteen la otra mejilla después
que una garra me ararló esta suerte.
Sólo la pido a Dios
que la guerra no me sea indiferente,
es un monstruo grande y pisa fuerte
toda la pobre inocencia de la gente.
Sólo le pido a Dios
que el engano no me sea indiferente
si un traidor puede más que unos cuantos,
que ecos cuantos no lo ohriden facilmente.
Sólo le pido a Dios
que el futuro no me sea indiferente,
desahuciado está el que liene que marchar
a vivir una cultura diferente.
Sólo le pido a Dios
que la guerra no me sea indiferente,
es un monstruo grande y pisa fuerte
toda la pobre inocencia de la gente.
Leon Gieco

I only ask of God
that pain not be indifferent to me,
that dry bony death not find me empty and alone
without having done enough.
I only ask of God
that I won't be indifferent to the injustice
that they won't slap my other cheek,
after a claw (or talon) has scratched this destiny (luck) of mine.
I only ask of God
that I am not indifferent to the battle,
it's a big monster and it walks hardly on
all the poor innocence of people.
I only ask of God
that I am not indifferent to deceit,
if a traitor can do more than a bunch of people,
then let not those people forget him easily.
I only ask of God
that I am not indifferent to the future,
hopeless is he who has to go away
to live a different culture.
I only ask of God
that I am not indifferent to the battle,
it's a big monster and it walks hardly on
all the poor innocence of people.
Leon Gieco

The question now remains as to what this song is doing at the end of our study. It appears to bear no relation whatsoever to what went before. I will try to explain. In the world and in the country we live in today, I have had the fortunate opportunity to talk and hold discussions with many people from different generations, be it in lectures, lessons, events, be it also in the close contact with 20-year old youths, students of business administration at the USP university, with whom I spend most of my time. It is a new generation, it is the new target public for agribusiness. This generation does not know what inflation is, or cassette tapes, floppy disks, the military dictatorship, and they never saw Ayrton Senna or Bebeto and Romário bringing home the World Cup in 1994. It is this first that, for me, is very different.

In my to-ing and fro-ing, with all of these different publics, what worries me most today, what I see as being the worst problem for society, is indifference. Everybody is indifferent to everything. Things happen and people, in their comfort zones, keep letting them happen. They do not move, they do not try to make a difference. These lyrics by Leon Gieco are entirely dedicated to combating indifference. I invite you to read the lyrics again.

We began in October 2009 by talking with the citrus farming community, which at that time had a new sectorial organization for the industry, until then seriously lacking the means to deal with the challenges facing the productive chains of Brazilian agribusiness. There was considerable wariness. I was a staunch critic of the disarticulation of citriculture, in interviews, lectures and classes. And in articles in the easily accessible news publications Folha, Estado and Valor.

Our talks were aimed at updating the 2003 study for mapping and quantification of the productive chain for citrus products. Yet another of our studies looked at the mapping and planning of productive chains. I always hold on to the scientific outlook of the people affiliated with the University of São Paulo: to research, gather data, perform analyses and bring answers. But more than this: to publish articles, books and other forms of divulging information, fulfilling the basic role of the university, which is to carry out research and publish results promoting improvement in our competitive environment and developing young talent.

The study started out lukewarm. Members of CitrusBR had different views about what they wanted, about what was to be done. We gathered the traditional data on supplies, on mediating companies, service providers, agriculture and the basics for the industry. At first the e-mails went unanswered and it was hard to arrange interviews and visits to the industries.

To our surprise, the meetings with the four companies began to take shape. The people came to like and support the idea. Repeating the words of one of the partners. "Professor, we have decided to add traction to your study". I will never forget it. There were more participants, a greater presence of the technicians, and more information poured in. Other companies in the productive chain embraced the cause and decided to give more information, as was the case for TetraPak, to mention just one. What was set to end in February 2010 went on to October 2010. And let their be traction...

When all is said and done, I am happy that our study has served as a catalyst for the citrus farming industry, so criticized for its closed nature, to gain confidence and trust regarding the need to publish its information on an aggregated basis. An international auditing firm was hired and given access to the individual data, which it consolidated applying total confidentiality. We also prefer this solution for such sensitive data. It is all here; published and analyzed.

The indifference seems to have been broken. The reader has seen the amount of information and the analyses we have carried out. We have been shown a world we had never seen before. The European world, the American world and the world within the manufacturing plants, among others. We left a number of meetings in São Paulo, Matão, Indiantown, Modena, Nice and other places literally exhausted by the volume of information and data. A doubt was always left hanging among the researchers. Is it possible that they might cut this out at the end of the study? They didn't.

The future agenda for the sector is complex and known. Here we present some of the opinions discussed in these two months.

- to strengthen the representative associations to enrich the debate in favor of uniting the links in the productive chain;
- to form an association along the lines of the Orplana (by the Consecana), which represents the current citriculture organizations;
- to construct reliable technical solutions for the purchase of oranges as "soluble solids";
- to disseminate the best practices for agricultural management aimed at increasing the productivity and competitiveness of the chain;
- to institute covenants with the agribusiness universities, with the aim of creating excellent technical and economic databases;
- to publicize, with greater transparency, information of relevance to the sector;
- to promote campaigns for the growth in Brazilian consumption of oranges and orange juice;
- to incessantly seek the reduction of custom duties in the import markets for Brazilian juice; and
- to work with the government agencies in order to obtain funding to support citriculture.

The changes seen throughout the productive chain have the same origin: an understanding that the end consumer does not want to and will no longer pay for the inefficiencies in the chain of supply. The demands of this new order have imposed challenges that cannot be met under the pretext of an isolated and static system. Only coordination of the chain as a whole and the incessant quest for efficiency and low costs will be able to boost the performance of all the links that make up the chain.

In my view, citriculture, which we used to criticize so heavily for its lack of coordination and articulation, has turned the tables, broken down the indifference and taken the talks to a new level. Let it be a first step toward the joint construction of a new phase in this productive chain, marked by more harmonious relationships, transparency, integrated work, asset sharing, combat against production costs and other threats, so that the professionals can focus their attention on the more serious problems, which are the drop in the consumption of juice and the changes in consumer habits. These are the main problems. We need to solve the other issues quickly in order to focus our attention on these points so that citriculture can bring São Paulo and Brazil a further US$ 60 billion to US$ 100 billion in the next 50 years.

Once again we have fulfilled our role as educators. All in all, almost ten young graduate students and post-graduates from USP and one from UNESP took part in this survey, received grants and will be encharged with scientific work, dissertations and theses. I will not name all of the researchers here, but I would like to underline the fundamental role played by the PhD student Vinícius Trombin. These young people have developed a taste for research and will continue our work in the future, just like me, when in the third year of Agronomic Engineering at ESALQ (in the far gone days of 1989) I was invited to participate in a survey similar to this one by my teacher Evaristo Marzabal Neves and decided to continue. Congratulations to the four members and to CitrusBR for having aided and supported this study, which, today, represents further patrimony in Brazilian knowledge. Our special thanks to Christian, the Chairman of that institution.

We would also like to thank a number of people and organizations who were of fundamental importance to this study.

a. All of the private companies interviewed, including suppliers, packing houses, packaging companies, service providers, producers, financial agents and others, who, through the generous offer of their data made it possible for us to conduct a well-rounded research project.
b. We would also like to thank the personalities from the sector, such as Mr. Antonio Ambrosio Amaro, a veritable encyclopedia on citriculture, always willing to share data and information and to cooperate with our research.
c. Thank you to the friends from the GCONCI (Citrus Farming Consultants Group) who were always available and who, for 14 years, have been providing services of extreme relevance to citriculture.
d. Thank you to the friends from GTACC (Technical Group for Assistance and Consulting in Citrus Farming) who were always collaborative.
e. Thank you to the staff from the Sylvio Moreira APTA center, here representing all of the research agencies for the sector.
f. Thanks to Margarete Boteon and the personnel from CEPEA, all of whom played a fundamental role in tracking costs and prices.
g. Thanks to Mauricio Mendes and to AGRA FNP, also for their constant cooperation.

Without these professionals and many others who are not mentioned here, it would not have been possible to conclude this study.

Citrus farming productive chain – let's get to work. Brazil is waiting for its dollars.

Marcos Fava Neves
Chair of Planning at the University of São Paulo
Head of the Department of Administration at FEA/USP, for the Campus in Ribeirão Preto.

Markestrat

Markestrat, Center for Research and Development in Marketing and Strategy, is an organization founded by doctorate and PhD graduates in Business Administration, who have graduated from the Faculty of Economics, Administration and Accounting (FEA) at the University of São Paulo (USP). The group was founded in 2004 by the teacher Marcos Fava Neves, whose aim it was to carry out studies and projects in Marketing and Strategy within a variety of sectors of the economy. Markestrat focuses its work on the analysis, planning and implementation of strategies for market-oriented companies with a focus on productive networks.

The global relationship network at the Center for Research and Development is made up of professionals, companies, universities and related research and development centers aimed at offering its services directed at study and research, further education and extension projects.

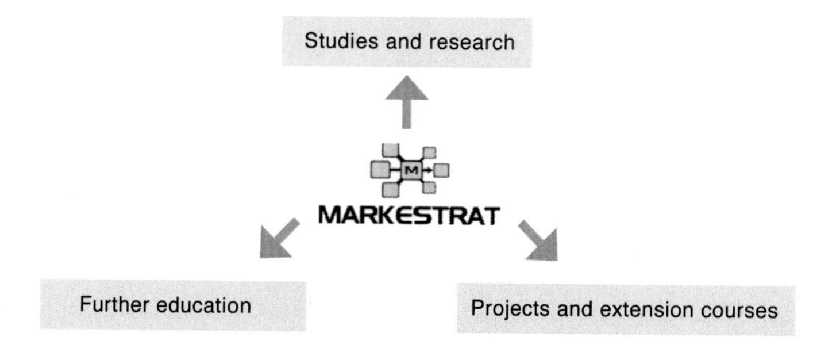

The role that Markestrat seeks to play in society is that of 'developing and applying knowledge on the planning and management of strategy and marketing in productive networks, aiming to increase the competitiveness of the companies via interaction between research, teaching and extension course'.

The work methodology used by Markestrat is based on the systematic analysis of businesses, with a special focus on the interfaces and networks created between the various sectors. This methodology is also backed up by two principles: analysis of the issues surrounding the decision process in the organizations and a concern for bringing the existing knowledge closer to the reality and needs of the market.

Here is the background of the authors of this survey

The history of the authors relating to citriculture goes back a long way. Here we indicate some of the most important events in this trajectory.

We start in the year of 1989, when Marcos Fava Neves, currently a professor at the University of São Paulo, began his work on the economy of citrus farming as an intern for the teacher Evaristo Marzabal Neves in the Department of Economics ESALQ. In 1995, Marcos presented his Master's thesis at the University of São Paulo talking about the economics of transaction and contract costs in citriculture. Later, in 2005, he supervised the Master's dissertation of Frederico Fonseca Lopes regarding the mapping of the productive chain for oranges, a work that gave rise to the book Estratégias para a Laranja no Brasil (Strategies for Oranges in Brazil), published by Atlas in 2005 with 2,406 copies sold. In 2007, Vinícius Gustavo Trombin presented his Master's thesis, also at the University of São Paulo, talking about a lengthy survey, supported by Codevasf, which analyzed the viability of transplanting a part of the citriculture in the State of São Paulo to the irrigated agricultural center of Petrolina-Juazeiro, in the Northeast of Brazil. In that same year, another work was published by Atlas under the title of Caminhos da Citricultura (Paths of Citriculture), with more than 2,000 copies sold, portraying a profound diagnosis of the sector and alternatives for its development. This latter work was written with the participation of the illustrious authors Antonio Ambrosio Amaro, Evaristo Marzabal Neves and Marcos Sawaya Jank.

Such theses and books are the instruments used by the authors to disseminate knowledge on citriculture and to leave it for posterity. However, they are not the only means of doing so. On this journey with citriculture, they have given a range of lectures, be it in the consecrated citriculture week in Cordeirópolis or at regional seminars in places where the orange has been gaining ground, such as in the State of Bahia, in the State of Sergipe and in the Southern regions of this country. In this way they take, often unprecedented, information to the different agents in the sector. It is also in this way that, by studying the sector, they learn from those that have been dedicating themselves for a much longer time to the quest for a stronger citriculture.

This Executive Summary is a new work by these authors who now share this authorship with other young initiates in citriculture. So that this story does not terminate here, this work is also to be released, in a more complete version for a major Brazilian publishing company. It is in this way that the history of the orange is being told, in a complete and transparent format. How good it would be to see something similar in other sectors of the Brazilian economy. Congratulations to citriculture!

Appendix

Average prices for orange juice on the New York Stock Exchange (US$ per pound weight of solids).

Harvest	July	August	September	October	November	December	January	February	March
1967/68	37.76	36.01	36.78	43.59	52.88	58.83	55.25	51.22	53.38
1968/69	48.15	56.88	62.68	68.41	59.02	54.42	65.10	65.61	58.67
1969/70	49.25	46.37	45.78	45.46	42.50	41.87	50.05	42.23	36.85
1970/71	37.16	37.80	36.83	35.14	35.18	36.54	36.86	44.58	46.96
1971/72	60.60	60.17	57.31	62.03	65.18	61.42	57.92	56.30	53.13
1972/73	53.82	54.37	53.03	48.10	48.25	46.48	43.77	43.32	44.06
1973/74	47.38	48.52	48.85	53.20	56.55	56.11	52.51	51.54	47.15
1974/75	50.10	51.75	52.89	53.30	55.13	53.11	48.45	47.51	48.43
197576	56.17	59.33	61.76	61.69	62.00	59.67	59.24	62.20	61.60
1976/77	53.54	50.38	49.74	48.54	47.50	43.52	48.64	72.12	77.77
1977/78	104.22	116.41	121.38	125.41	128.59	110.63	108.50	121.03	119.01
1978/79	122.53	123.22	119.36	121.66	119.89	114.10	118.57	113.06	103.79
1979/80	100.46	106.97	107.47	106.23	101.20	97.35	91.25	85.43	95.31
1980/81	87.75	91.71	97.40	94.10	89.94	82.94	104.25	137.13	135.59
1981/82	126.79	126.32	127.42	121.20	120.74	122.75	138.29	133.39	120.02
1982/83	124.93	129.31	127.22	125.06	125.21	123.95	111.56	107.57	113.41
1983/84	118.61	118.70	120.83	123.89	128.57	126.26	149.66	161.28	168.30
1984/85	171.39	171.99	177.63	169.99	167.10	161.96	166.13	170.16	163.71
1985/86	136.98	134.08	135.13	121.03	113.47	116.02	96.89	86.75	88.33
1986/87	103.03	101.49	103.62	112.01	121.71	126.84	122.24	123.41	132.68
1987/88	129.32	129.51	134.64	142.53	163.10	167.65	169.99	168.05	166.37
1988/89	190.06	193.35	184.98	185.24	177.91	164.20	148.08	138.39	149.22
1989/90	166.48	158.86	148.29	133.07	128.97	135.61	191.30	197.74	192.27
1990/91	183.34	172.24	144.56	123.08	112.72	108.43	118.19	117.07	115.64
1991/92	118.65	118.09	120.64	151.01	168.77	160.40	149.59	141.87	143.36
1992/93	121.78	112.92	114.27	101.12	95.52	94.56	78.91	69.11	78.46
1993/94	119.03	118.71	122.53	119.39	104.81	105.97	108.47	105.83	109.50
1994/95	89.99	94.12	90.25	100.12	108.99	111.21	103.33	102.68	100.98
1995/96	97.82	105.00	111.61	115.96	123.27	120.90	117.93	124.16	132.78
1996/97	116.40	117.20	110.14	111.50	101.59	88.70	83.56	80.36	82.98
1997/98	74.86	72.21	69.99	69.82	78.02	84.11	90.98	97.67	105.94
1998/99	104.01	110.18	108.18	115.24	117.72	108.57	99.66	93.00	83.48
1999/00	80.16	92.55	92.97	88.52	94.85	93.19	84.37	84.66	84.82
2000/01	79.65	74.07	71.42	70.03	73.99	80.42	76.01	75.69	74.80
2001/02	81.36	77.68	80.81	85.26	93.76	91.67	89.41	89.64	92.68
2002/03	95.42	100.93	100.30	95.14	100.53	97.36	92.04	87.08	84.68
2003/04	81.31	78.66	77.17	70.77	70.11	67.01	62.95	60.97	61.21
2004/05	66.74	67.27	79.99	82.49	74.99	83.46	82.11	85.02	94.84
2005/06	99.88	91.59	95.72	108.03	119.95	125.06	123.07	130.18	139.94
2006/07	163.04	177.49	174.78	184.62	197.72	201.23	200.33	195.62	199.98
2007/08	133.10	129.60	125.06	142.53	136.43	144.36	136.92	128.23	118.80
2008/09	121.63	103.60	94.76	81.41	79.81	73.82	74.40	69.25	73.72
2009/10	93.66	97.76	93.61	107.96	113.27	128.83	137.47	137.38	146.30

Source: Statistical Annuals from Cacex, Bank of Brazil and Siscomex.

April	May	June	Harvest average	Deduction of American import duty	Stock exchange average minus import duty	Equivalence for price of orange juice minus import duty (US$ per ton 66° Brix)
55.30	55.31	49.40	48.81	-34.01	14.80	215
55.78	51.77	51.12	58.13	-34.01	24.12	351
39.50	38.35	35.01	42.77	-34.01	8.76	127
52.12	58.46	63.35	43.42	-34.01	9.40	137
49.68	53.40	52.63	57.48	-34.01	23.47	342
43.07	43.62	43.81	47.14	-34.01	13.13	191
47.34	47.93	48.81	50.50	-34.01	15.93	232
48.09	51.35	52.91	51.09	-34.01	17.08	248
60.15	58.50	55.07	59.78	-34.01	25.77	375
79.48	83.69	96.48	62.62	-34.01	28.61	416
117.11	112.12	118.09	116.88	-34.01	82.87	1,206
106.21	103.13	97.44	113.58	-34.01	79.57	1,158
89.13	88.75	86.65	96.35	-34.01	62.34	907
143.35	140.25	134.25	111.56	-34.01	77.55	1,128
115.15	117.14	116.05	123.77	-34.01	89.76	1,306
114.37	116.62	117.01	119.69	-34.01	85.68	1,247
179.80	184.26	178.17	146.53	-34.01	112.52	1,637
157.46	151.54	142.57	164.30	-34.01	130.29	1,896
93.13	97.89	101.06	110.06	-34.01	76.05	1,107
133.52	133.58	132.62	120.56	-34.01	86.55	1,259
170.20	169.36	176.80	157.29	-34.01	123.28	1,794
171.90	186.42	180.65	172.53	-34.01	138.52	2,016
196.04	194.95	186.45	169.17	-34.03	135.14	1,966
115.07	119.10	116.31	128.81	-34.04	94.77	1,379
136.06	135.67	129.04	139.43	-34.04	105.39	1,533
90.65	102.46	112.91	97.72	-34.04	63.68	927
102.21	96.50	92.44	108.78	-34.04	74.74	1,088
107.01	104.65	100.90	101.19	-33.62	67.57	983
132.07	123.23	122.17	118.91	-32.75	86.16	1,254
75.13	78.64	75.95	93.51	-31.89	61.63	897
97.07	109.96	103.73	87.86	-31.04	56.82	827
84.47	85.42	89.23	99.93	-30.18	69.76	1,015
82.49	81.77	84.44	87.07	-29.31	57.76	840
74.25	78.33	77.02	75.47	-28.89	46.58	678
89.61	91.17	91.40	87.87	-28.89	58.98	858
85.49	85.74	85.30	92.50	-28.89	63.61	926
59.45	56.11	57.66	66.95	-28.89	38.06	554
95.29	93.71	96.24	83.51	-28.89	54.62	795
144.84	155.09	158.23	124.30	-28.89	95.41	1,388
171.73	165.03	138.36	180.83	-28.89	151.94	2,211
115.73	112.30	112.00	127.92	-28.89	99.03	1,441
80.55	90.74	81.95	85.47	-28.89	56.58	823
133.12	140.29	141.05	122.56	-28.89	93.67	1,363

Conversion table

1 hectare	= 10,000 m^2
1 acre	= 0.40469 hectares
1 90 lb box	= 40.8 kg
1 pound weight	= 0.453593 kg
1 gallon	= 3.785 liters
1 ton 66° Brix	= 5,295.5 liters of ready-to-drink juice

Metric conversions

1 40.8 kg box of oranges	= 90 pounds of fruit
1 metric ton	= 2,204.60 pounds
1 pound	= 0.454 kg
1 kg	= 2.2046 pounds
1 gallon	= 3.785 liters
1 liter	= 0.2641 gallons
1 hectare	= 2.47 acres
1 acre	= 0.405 hectares

Content of soluble solids in 1 metric ton of FCOJ

1 ton of FCOJ at 65.0° Brix	= 1,433 pounds of solids
1 ton of FCOJ at 66.0° Brix	= 1,455 pounds of solids
1 ton of FCOJ at 67.0° Brix	= 1,477 pounds of solids

Volume occupied by 1 ton of FCOJ

Brix bands	Gallons	Liters
65.0° Brix	204	772
66.0° Brix	200	757
42.0° Brix	350	1,326
11.8° Brix	1,414	5,354

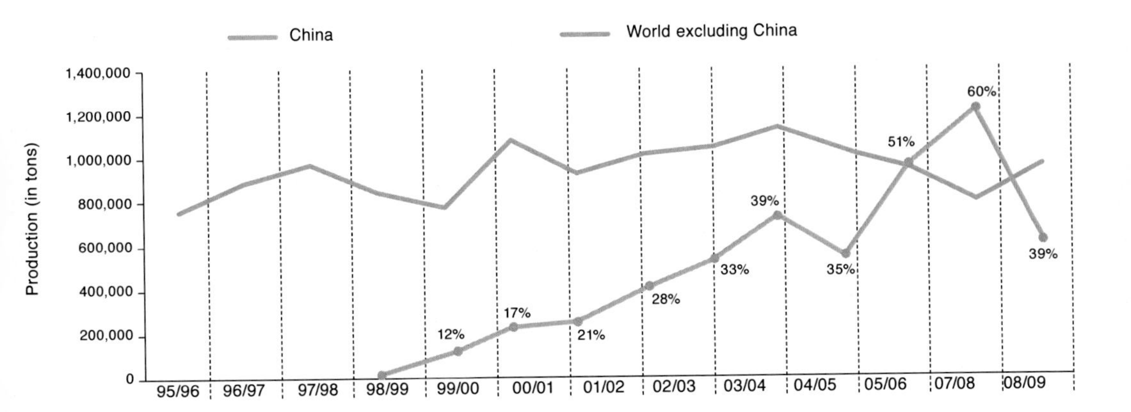

World production of apple juice

Source: prepared by Markestrat based on data from Cepea.

Glossary

BACEN	**Brazilian Central Bank.**
BRIC	Term invented by Jim O'Neal, economist for Goldman & Sachs, to define the emerging nations: Brazil, Russia, India and China.
Brix	Brix (symbol °Bx) is a numerical scale that measures the quantity of soluble solids in a saccharose solution. The Brix scale is used in the food industry to measure the approximate quantity of sugars in fruit juices.
Capex	Acronym for the expression Capital Expenditure, which designates the amount of money spent on the acquisition (or introduction of improvements) in capital goods by a given company.
CEPEA	Center for Advanced Studies in Applied Economics ESALQ/USP.
CitrusBr	National Association of Citrus Juice Exporters.
COFINS TAX	The Social Security Funding Contribution (COFINS) is a federal tax applied to the gross revenues of companies in general, destined for the funding of social security
Commodity	This terms means merchandise and is used in commercial transactions for products of primary origin in the commodities exchanges. Plural: commodities.
CPR	Rural Product Bill: This is a declaratory and exchange bond that allows the producer to receive credit, using their production as a guarantee.
DR	Rural Trade Bill: Bond representing credit for term sales made by the producer or rural cooperative
EGF	Federal Government Loans. This is a line of credit to finance the stocking of agricultural products for the Minimum Price Guarantee Policy (MPGP) for future sale under better market conditions.
IPE	Individual Protection Equipment
FAO	United Nations Food and Agriculture Organization.
FCOJ	Frozen Concentrated Orange Juice.
FDOC	Florida Department of Citrus.
IBGE	Brazilian Institute of Geography and Statistics.
ICE	Intercontinental Commodity Exchange.
ICMS TAX	Brazilian Goods and Services Tax relating to the Circulation of Goods and the Provision of Services in both Interstate and Intercity Transport, as well as Communication. The ICMS is a state tax.
IEA	Institute of Agricultural Economics for the São Paulo State Department of Agriculture.
INPC	National Consumer Price Index.
IPC	Consumer Price Index.
IPI	The IPI (Industrialized Goods Tax) is a federal tax.
Physical Market	Market on which the buying and selling transactions for the physical asset/object are made according to terms accepted both by the buyer and by the seller.

Futures Market	Market on which operations are carried out involving standardized lots of commodities or financial assets for settlement on pre-established dates.
Spot Market	Physical cash market.
NFC	Not From Concentrate
NPR	Rural Promissory Note.
NR 31	Regulatory Norm 31, relating to health and safety at work in agriculture, animal husbandry, silviculture, forestry exploitation and aquiculture.
Options	Contracts that involve the establishing of rights and obligations on certain bonds, with pre-established terms and conditions.
Hedging operation	Protection against price variation via the purchase or sale of futures contracts or options.
Packing House	Fruit processing warehouse.
GDP	Gross Domestic Product: Represents the sum (in monetary values) of all the finished goods and services produced within a given region.
PIS	Social Integration Program: This is a social tax contribution paid by corporate entities with the aim of funding the payment of unemployment benefits and bonuses for workers that earn up to twice the minimum monthly wage.
Players	Market operators.
Pools	Groups of producers that come together to jointly negotiate their production.
SECEX/ MDC	Department of Foreign Trade at the Ministry of Foreign Development, Trade and Industry.
Tank Farm	Tank for bulk storage of juice.
USDA	United States Department of Agriculture.

The study coordinators

Marcos Fava Neves

Agronomic Engineer graduated from ESALQ/USP in 1991, PhD in Administration (Industrial Leasing Strategies, FEA/USP, 1995) and Doctorate in Administration (Planning of Channels of Distribution, FEA/USP, 1999). Post-Graduate in European Agribusiness & Marketing in France (1995) and in Channels (Networks) of Distribution in Holland (1998/1999). Livre-Docente (Qualified to Teach) (Strategic Planning and Management Driven by Demand (2004). Coordinator of PENSA – the Agribusiness Program run by USP (strategic planning for companies and productive systems from 2005 to 2007) he is the creator of Markestrat (Center for Research and Development in Marketing and Strategy), having carried out 70 projects and given 300 lectures in Brazil. He is the author/co-author and organizer of 25 books in Brazil, Argentina, the United States, South Africa, Uruguay and the European Union. His work is characterized by the proposal of methods (frameworks) for the solving of problems in business and productive chains and for international insertion. He has taken part in more than 60 congresses overseas and has given more than 120 international lectures. He has had 70 articles published in international periodicals and annals of scientific meetings, as well as 45 articles published in indexed magazines in Brazil. He has sat on more than 150 panels in Brazil and provided guidance for 20 Master's and PhD works at USP. He is specialized in strategic planning and management. Feature writer for the China Daily newspaper, from Peking, China, and the Folha de São Paulo newspaper, he also wrote about two cases for the University of Harvard in 2009 and 2010.

He is the author/co-author and organizer of 25 books:

- Published by Routledge (USA):
  - "Marketing Methods to Improve Company Strategy" (2010)
  - "Demand Driven Strategic Planning" (on the printing press, 2011)
- Published by Editora Monteverde (Uruguay):
  - "El Futuro de Los Foodstuffs y Uruguay" (2010)
- Published in South Africa:
  - "The Future of Food: Messages to South Africa" (2010)
- Published by the University of Buenos Aires (Argentina):
  - "Agronegócios en Argentina y Brasil" (2007)
- Published by Editora Atlas (Brazil):
  - "Integrated Agriculture" (2010)
  - "Strategies for Sugarcane in Brazil" (2010)
  - "Strategic Planning for Events" (2008)
  - "Competitive Resale in Agribusiness" (2008)
  - "Agribusiness and Sustainable Development" (2007)
  - "Paths for Citriculture" (2007)

- "Strategies for Milk in Brazil" (2006)
- "Strategies for Oranges in Brazil" (2005)
- "Strategic Planning and Management for Marketing" (2005)
- "Sales Administration" (2005)
- "Strategies for Wheat in Brazil" (2004)
- "Marketing and Strategy in Agribusiness and Foodstuffs" (2002)
- "Marketing & Export" (2001)
- "Marketing in the New Economy" (2001)

- Published by Editora Malcron Books (Brazil):
  - "Sales Planning" (2007)
- Published by Editora Saraiva (Brazil):
  - "Agribusiness in Brazil" (2005)
- Published by Thomson Lcarning (Pioneira, Brazil):
  - "Economics and Business Management for Agricultural Foodstuffs" (2000)
  - "Foodstuffs, New Times and Concepts in Business Management" (2000)
  - "Case Studies in Agribusiness" (1998)
  - "European Agribusiness" (1996)
- Published by SEBRAE – SP (Brazil):
  - "Strategic Planning and Management of the Agroindustrial System for Milk" (2008)

Main projects

- Project for Analyzing the Attractiveness of 10 Local production Arrangements in the State of Minas Gerais for Sebrae MG in 2010.
- Orange Chain Mapping Project for CitrusBR in 2010.
- Strategic Plan for Brazilian Fruit Farming, for IBRAF/APEX in 2010.
- Strategic Plan for Brazilian Cattle, for ABCZ/APEX in 2010.
- Planning for the Dairy Sector in Uruguay (INIA – CRI Lechero) in 2010.
- Sugarcane Chain Mapping Project for UNICA in 2009.
- Analysis of International investments in the Sugarcane Chain, for UNCTAD/ONU in 2009.
- Q-Pork Chains Project (Transnational Productive Chains for Pork), for the European Union (2006-2010).
- Planning for Renk Zanini and other companies within the Biagi family in 2008.
- Supply Chain Analysis for the Zilor Group (Zillo Lorenzetti) in 2008.
- Analysis of the Brazilian Citrus products Chain for FAO/ONU in 2007.
- New Remuneration Project for Fundecitrus in 2007.
- Planning Project for the Wheat Board in Uruguay in 2007.
- Strategic Planning and Management for the Milk Chain in São Paulo, for Sebrae – 2007
- Strategic Plan for the Orange Chain in Brazil, in 2007
- Planning for the implantation of 10 businesses in the São Francisco valley, contracted by Codevasf, 2007-2008.
- Strategic Outlook for the Branco Peres Sugar and Alcohol Group in 2007.

- Analysis of scenarios for the Zillo Lorenzetti sugar group in 2005.
- Planning of Channels of Distribution for Basf in 2004-2007.
- Planning for the implantation of citriculture in the Petrolina Juazeiro region – 2006.
- Strategic Planning and management for the Organização Laranja Brasil (Brazil Orange organization) in 2003.
- Strategic Planning and Management for the Wheat Chain in Brazil in 2003.
- Strategic Planning and Management for Lagoa da Serra from 1999 to 2006.
- Strategic Planning and Management for Netafim do Brasil from 2001 to 2004.
- Strategic Planning and Management for Wolf Seeds/Naterra in 2004/05.
- Joint venture project for Tigre in 2004.
- Planning of Channels of Distribution for Orsa Embalagens, in 2002.
- Analysis of the capture of value in the soy chain for Monsanto in 1998.
- Project for strategic creation/planning for Crystalsev, (sugar-alcohol) in 1997.
- Projects also conducted for the following companies: Vallée (veterinary products), Big/Real Supermarkets (Retail), Arby's (food service), Sanavita (functional foodstuffs), Boehringer (veterinary products), Illycafé (expresso coffee – Italian multinational), Fri-Ribe (animal feeds), J. Mãcedo Foodstuffs (Dona Benta), Nestlé (foodstuffs), Elanco (animal healthcare).
- Project and Coordination of 14 class groups for MBA courses in Marketing at Fundace since 2000 (classified by the Você S.A. magazine as the best in Brazil in 2003).
- Project and Coordination of 19 class groups for open and in company MBA courses in Agribusiness at Fundace.
- Project Coordinator for Fundace since 1996 and Chairman of the Board of Curators from 2005 to 2007.
- Participated in the Global Food Network project to set up transnational chains between Mercosur and the European Union, run by the European Union from 2002-2005.
- A number of his projects have been financed by the UNO, FAO and UNCTAD and he has been a grantee of Fapesp, CAPES, CNPq and USP since 1989.

Vinicius Gustavo Trombin

Business Administrator from Federal University of Uberlândia in 2004, MSc in business from University of São Paulo in 2007 and PhD candidate at FEA-USP in Scenario Analysis and Strategic Planning. Prior to joining Markestrat (Center for Research and Development in Marketing and Strategy), he worked to the largest wholesales in Brazil, Martins and Makro. In 2007 as a researcher at Markestrat, he develop a method of viability analysis for the implantation of a productive chain in a new place, as member of the Markestrat team that analyzed the viability to transplant part of the citrus chain from Sao Paulo State to Bahia and Pernambuco, northeast states in Brazil. He has worked on different projects advising the top management on strategic planning of value chains and how increase competitiveness in agri-food chains. His academic career as professor began in 2007, being honored professor in all undergraduate class he taught. Also he teaches business management in MBA courses, PECEGE – USP and

FAAP University. His researches and ideas are in many articles published in international periodicals, annals of scientific meetings and magazines in Brazil.

He is the author/co-author and organizer of five books:

- Published by Wageningen Academic Publishers:
  - "Food and Fuel: the example of Brazil" (2011)
- Published by Editora LUC (Brazil):
  - "Ethanol and Bioelectricity sugarcane in the Future of the Energy Matrix" (2010).
- Published by Editora Atlas (Brazil):
  - "Integrated Agriculture" (2010)
  - "Agribusiness and Sustainable Development" (2007)
  - "Paths for Citriculture" (2007)

Main Projects

- Strategic Plan for ABCZ in 2011.
- Project for Analyzing the Attractiveness of 10 Local production Arrangements in the State of Minas Gerais for Sebrae MG in 2010.
- Orange Chain Mapping Project for CitrusBR in 2010.
- Strategic Plan for Brazilian Fruit Farming, for IBRAF/APEX in 2010.
- Strategic Plan for Brazilian Cattle, for ABCZ/APEX in 2010.
- Strategic Plan for Plumrose/Venezuela in 2009.
- Provide a trainning in Strategic Plan for CRV Lagoa in 2009.
- Development of alternative production in Public Irrigation Projects for IFC in 2009.
- Implementation of Customer Relationship Management in channels of Distribution for Basf in 2008.
- Sugarcane Chain Mapping Project for UNICA in 2009.
- Strategic Analysis for Citrofruit/Mexico in 2008.
- Financial viability analysis in Irrigation Public Areas for Queiroz Galvão Group in 2007.
- New Remuneration Project for Fundecitrus, in 2007
- Strategic Plan for the Orange Chain in Brazil in 2007.
- Planning for the implantation of 10 businesses in the São Francisco valley, contracted by Codevasf, 2007-2008.
- Planning for the implantation of citriculture in the Petrolina Juazeiro region – 2006.

List of important websites for further information

ANDA	National Association of Fertilizers Distributors.	http://www.anda.org.br
ANP	National Agency of Petroleum, Natural Gas and Biofuels.	http://www.anp.gov.br
BACEN	Central Bank of Brazil	http://www.bcb.gov.br
BB/ CACEX	Bank of Brazil. Foreign Trade Department of the Bank of Brazil.	http://www.bb.com.br
CAGED	General Register of Employed and Unemployed.	http://www.caged.gov.br
CEPEA	Center for Advanced Studies on Applied Economics.	http://www.cepea.esalq.usp.br
CitrusBR	Brazilian Association of Citrus Exporters.	http://www.citrusbr.com
CONAB	National Supply Company.	http://www.conab.gov.br
Dhöler	Natural Food & Beverage Ingredients	http://www.doehler.com/
FAO	Food and Agriculture Organization	http://www.fao.org
FDOC	Florida Department of Citrus	http://www.floridajuice.com
FUNDECITRUS	Citriculture Defense Fund	http://www.fundecitrus.com.br
GV Agro Pesquisa	Getulio Vargas Foundation Agribusiness Center	http://www.eesp.fgv.br
IBGE	Brazilian Institute of Geography and Statist	http://www.ibge.gov.br
ICE Futures	Intercontinental Exchange	http://www.theice.com
IFAS	Institute of Food and Agricultural Sciences	http://www.ifas.ufl.edu
IAC	Agronomical Institute of Campinas	http://www.iac.br
IGD	IGD Retail Analysis	http://www.igd.com/analysis
IPEADATA	Institute of Applied Economic Research	http://www.ipeadata.gov.br
MIDC	Ministry of Development, Industry and Foreign Trade	http://www.mdic.gov.br
MARM	Ministry of Environment and Rural and Marine Affairs of Spain	http://www.marm.es
MySupermarket	MySupermarket	www.mysupermarket.co.uk
Nielsen	The Nielsen Company	http://www.nielsen.com
UN	United Nations	http://www.un.org
Planet Retail	Global Retail Intelligence	http://www1.planetretail.net/
RAIS	Annual Report of Social Information	http://www.rais.gov.br
SECEX/MIDC	Secretary of Foreign Trade	http://aliceweb.desenvolvimento.gov.br/
SINDAG	National Union of Crop Protection Industry	http://www.sindag.com.br
SISCOMEX	Integrated Foreign Trade System	http://www.receita.fazenda.gov.br/aduana/siscomex/siscomex.htm.
USDA	United States Department of Agriculture	http://www.usda.gov
WORLD BANK	World Bank Group	http://www.worldbank.org

Further reading

AGRAFNP, 2010. Informa Economics FNP. AGRIANUAL. Publication in portuguese. São Paulo, 2010. 523 p.

Etzel, M.J., Walker, B.J., Stanton, W.J., 2001. Marketing. Makron Books.

Florida Department of Citrus, 2010. Economic and Market Research Department. Citrus Reference Book. Flórida, USA, 102 p.

Hawkins, D.I., Best, R., Mothersbaugh, D.L., 2007. Comportamento Do Consumidor: Construindo a Estrategia de Marketing. (Publication in Portuguese language). 10ª (Campus). Rio de Janeiro: Campus.

Neves, M.F., 1999. The relationship of orange growers and fruit juice industry: an overview of Brazil. Fruit Processing, International Journal for the Fruit Processing and Juice Producing Industry 9(4): 121-124.

Neves, M.F., 2008. Método para planejamento e gestão estratégica de sistemas agroindustriais (GESIS). Revista de Administração, São Paulo 43: 331-343.

Neves, M.F., 2011. The future of food business – the facts, the impacts, the acts. Editora World Scientific Publishing Co. Pte. Ltd., Cingapura, Brazil, 173 pp.

Neves, M.F., Campos, E.M., Lopes, F.F., 2005. Opportunities for juice marketing channels. Fruit Processing, International Journal for the Fruit Processing and Juice Producing Industry 15(6): 378-381.

Neves, M.F., Castro, L.T., Gomes, C.C.M.P., 2002. Private labels in orange juice: in what should we think. Fruit Processing, International Journal for the Fruit Processing and Juice Producing Industry 12(10): 435-437.

Neves, M.F., Consoli, M.A., Neves, E.M., Jank, M.S., Lopes, F.F., Amaro, A.A., Trombin, V.G., 2007. Caminhos para a citricultura: uma agenda para manter a liderança Mundial. Editora Atlas, São Paulo, Brazil, 114pp.

Neves, M.F., Lopes, F.F. (eds.), 2005. Estratégias para a Laranja no Brasil. São Paulo: Editora Atlas, 224p.

Neves, M.F., Lopes, F.F., Santin, M.L., 2005. Opportunities for the Brazilian orange chain. Fruit Processing, International Journal for the Fruit Processing and Juice Producing Industry 15(3): 142-144.

Neves, M.F., Neves, E.M.,1999. The orange juice distribution channels: some characteristics, opportunities and threats. Italian Food & Beverage Technology 18: 15-28. Also published in Fruit Processing, International Journal for the Fruit Processing and Juice Producing Industry 9(8): 298-307.

Neves, M.F., Pinto, M.J.A., Conejero, M.A., Trombin, V.G., 2011. Food and fuel – the example of Brazil. Wageningen Academic Publishers, Wageningen, the Netherlands, 152pp.

Neves, M.F., Teixeira, L., Lopes, F.F., 2005. Juice consumer behavior in Brazil. Fruit Processing, International Journal for the Fruit Processing and Juice Producing Industry 15(2): 86-90.

Neves, M.F., Val, A.M., Marino, M.K., 2001. The orange network in Brazil. Fruit Processing, International Journal for the Fruit Processing and Juice Producing Industry 11(12): 486-490.

Neves, M.F., Zuurbier, P.J.P., Castro, L.T., Vlaar, P., 2000. The future competitiveness of the European Union food market. Italian Food & Beverage Technology 19: 13-20.

Pagliuca, L., Viana, M., Boteon, M., Inoue, K., Geraldini, F., Deleo, J.P.B., 2010. Sustentabilidade Econômica com HLB (Greening). Hortifruti Brasil. http://www.cepea.esalq.usp.br Acessed May 2010.

Printed in the United States
by Baker & Taylor Publisher Services